Praise f
Einstein's Unfinish

"Full of extraordinary ideas . . . Smolin's promised solution, when it comes, is just comprehensible enough for us to see that it's beautiful."
—*The Times* (London)

"Smolin is an extremely creative thinker who has been a leader in theoretical physics for many years. He is also a gifted writer who manages to translate his own insights about how science works into engaging language and compelling stories. . . . Smolin's description of how quantum mechanics works is both elegant and accessible."
—NPR

"Ambitious . . . upbeat and, finally, optimistic . . . Smolin is a lucid expositor."
—*Nature*

"As the latest entry into the conversation, Smolin's book feels the most immediate and personal. Here is no detached narrator, but an active participant in the fray who perceives the debate over the nature of reality in personal terms. . . . While the way forward remains elusive, Smolin and others who seek to illuminate how physics got to where it is today are at least making the quest for answers a bit less costly."
—*The Globe and Mail* (Toronto)

"Well-written and engaging."
—Sabine Hossenfelder, *Backreaction* (blog)

"Smolin offers a masterful exposition on the state of quantum physics, smoothly blending a history of the field with clear explanations, philosophical context, and an accessible introduction to fresh ideas. His narrative on how two competing perspectives on quantum behavior hardened into Bohr's counter-intuitive orthodoxy, is spellbinding."
—*Financial Times*

"Smolin is never less than an inventive and provocative thinker, as well as an engaging writer. . . . His explanations are especially lucid."

—Philip Ball, *Physics World*

"A tantalizing glimpse of the theoretical possibilities beyond Einstein's grasp."

—*Booklist* (starred review)

"Smolin elucidates complex science without equations . . . [and] demonstrates there isn't a thing in nature whose 'contemplation cannot be a route to a wordless sense of wonder and gratitude just to be a part of it all.'"

—*Publishers Weekly*

"The best explanation yet of what has yet to be explained."

—George Dyson, author of *Turing's Cathedral*

"This is a book about what it's like to treat life as an uncompromising search for the nature of reality. It is about physics, but also our historical moment. Do we essentially know everything already? Are we only cleaning up details and consolidating our powers, or is humanity still mostly ignorant? Smolin starts with a scrupulously fair telling of recent ideas in physics, and then accelerates into a vision of the future of physics. Sentences that are as simple as possible, like elementary particles, populate the closing argument. Read them slowly. They cut deep."

—Jaron Lanier, author of *Dawn of the New Everything* and *Ten Arguments for Deleting Your Social Media Accounts Right Now*

"Lee Smolin has written a superb and sweeping book. He takes us to Bohr, Bohm, Everett, and far beyond in a masterful assessment, then on to the struggle to go beyond quantum mechanics toward quantum gravity. *Einstein's Unfinished Revolution* is truly a fine work."

—Stuart Kauffman, author of *At Home in the Universe*

PENGUIN BOOKS

EINSTEIN'S UNFINISHED REVOLUTION

Lee Smolin has made influential contributions to the search for a unification of physics. He is a founding faculty member of the Perimeter Institute for Theoretical Physics. His previous books include *Time Reborn, The Trouble with Physics,* and *Three Roads to Quantum Gravity.*

ALSO BY LEE SMOLIN

The Singular Universe and the Reality of Time
(with Roberto Mangabeira Unger)

Time Reborn

The Trouble with Physics

Three Roads to Quantum Gravity

The Life of the Cosmos

EINSTEIN'S UNFINISHED REVOLUTION

The Search for What Lies Beyond the Quantum

Lee Smolin

Illustrations by Kaća Bradonjić

PENGUIN BOOKS

PENGUIN BOOKS
An imprint of Penguin Random House LLC
penguinrandomhouse.com

First published in the United States of America by Penguin Press,
an imprint of Penguin Random House LLC, 2019
Published in Penguin Books 2020

ISBN 9780143111160 (paperback)

THE LIBRARY OF CONGRESS HAS CATALOGED THE HARDCOVER EDITION AS FOLLOWS:
Names: Smolin, Lee, 1955– author. | Bradonjić, Kaća, illustrator.
Title: Einstein's unfinished revolution : the search for what lies beyond the quantum / Lee Smolin ; illustrations by Kaća Bradonjić.
Description: New York : Penguin Press, 2019. | Includes bibliographical references and index.
Identifiers: LCCN 2018045679 (print) | LCCN 2018060769 (ebook) | ISBN 9780698169135 (ebook) | ISBN 9781594206191 (hardcover)
Subjects: LCSH: Quantum theory. | Physics—Research.
Classification: LCC QC174.13 (ebook) | LCC QC174.13 .S6545 2019 (print) | DDC 530.12—dc23
LC record available at https://lccn.loc.gov/2018045679

Printed in the United States of America

DESIGNED BY AMANDA DEWEY

For Dina and Kai

All a musician can do is to get closer to the sources of nature, and so feel that he is in communion with the natural laws.

—JOHN COLTRANE

I can safely say that nobody understands quantum mechanics.

—RICHARD FEYNMAN

Contents

Preface

We human beings have always had a problem with the boundary between reality and fantasy. To explain the world to ourselves we make up stories and then, because we are good storytellers, we get infatuated by them and confuse our representations of the world with the world itself. This confusion afflicts scientists as much as laypeople; indeed, it affects us more, because we have such powerful stories in our tool kits.

As we go deeper into our understanding of the natural world, moving to smaller and more elementary phenomena, our successes impose barriers to further progress. To avoid getting stuck, we must balance our well-justified confidence in the power of established knowledge with an acute consciousness of just how hypothetical even our most successful hypotheses are. A hard lesson to learn is that our sensations are partly caused by reality, but are fully constructed by our brains to present the world to us in just the form we need to make our way in nature. Beyond those sensations, nature hovers, fundamentally mysterious and just at the edge of what we can know.

The most important features of nature, as we understand them now, were not perceived. The simplest general facts we know about the world—that matter is made of atoms, for example, or that the Earth is a spherical shell of rock surrounding a molten core and enveloped within a thin atmosphere, which moves, suspended in a near vacuum, as it orbits a natural thermonuclear reactor—these plain facts we learn just out of our cribs are the result of centuries of intense effort by scholars and scientists. Each of these facts arose as an almost crazy idea in conflict with a much more obvious and reasonable—but wrong—hypothesis.

To have a scientific mind is to respect the consensus facts, which are the resolution of generations of dispute, while maintaining an open mind about the still unknown. It helps to have a humble sense of the essential mystery of the world, for the aspects that are known become even more mysterious when we examine them further. The more we know, the more curious it all is. There is not a thing in nature so ordinary that its contemplation cannot be a route to a wordless sense of wonder and gratitude just to be a part of it all.

This spring morning the air coming through the open window carries fresh smells from the garden—but by what miracle does that happen? How are molecules wafted by a breeze turned by a nose into that happy scent? We see vivid colors, and we recall that there is a story about how different wavelengths of light excite different neurons. But how could the sensations of redness or blueness possibly be caused by different neurons being excited? What kinds of things are the sensations, the *qualia,* as the philosophers call them, of the different colors, or the different scents? In what way are scents different from colors, and why do they differ, if it is all electrical impulses in neurons? Who is the I that wakes and what is the universe that surrounds me when I open my eyes? The simplest facts about our existence and our relationship to the world are mysteries.

Let us tiptoe past the hard question of consciousness to simpler questions. As a scientist, I believe that is the best way to get somewhere. Let's start with one very basic question: What is matter? My son has left a rock on the table. I pick it up; its weight and shape fit comfortably in my hand—surely an ancient feeling.

But what is a rock?

We know what the rock looks like, what it feels like. But these are at least as much about us as they are about the rock. Little in a rock's feel or appearance gives a hint as to what, essentially, constitutes the existence—the rockness—of a rock. We know most of the rock is empty space in which atoms are arranged. The solidity and hardness of the rock is a construction of our minds, which integrate perceptions on scales very coarse compared to the sizes of the atoms.

Matter comes in many forms, some of which, like the rock, like the organic material woven into our blankets, sheets, and clothes, we know must be complex. So let's consider first a simpler form of matter: the water in our glass. What is it?

To our eyes and to our touch, water appears to be smooth, continuous. Until relatively recently, a bit more than a century ago, physicists thought that matter was entirely continuous. Early in the twentieth century, Albert Einstein showed that was wrong and that water is made of myriad atoms. In water, these are organized into triplets, bound together into molecules, each consisting of two hydrogen and an oxygen.

Yes, but what is an atom? It took less than a decade after Einstein for it to be understood that each atom is like a tiny solar system, with a nucleus in the center in place of the Sun and the planets represented by electrons.

So far so good, but then what is an electron? We know that electrons come in discrete units, each one carrying a certain quantity of mass and charge. An electron can have a location in space. It can move: when we first look it is here; when we look again it is there.

Beyond those attributes it is not easy to give a picture of what an electron is. It will take much of this book.

The best understanding of what rocks are, what water is, what molecules and atoms and electrons are, is expressed by the branch of science called *quantum physics*. But, as it seems everyone knows by now, that is a realm full of paradox and mystery. Quantum physics describes a world in which nothing has a stable existence: an atom or an electron may be a wave or a particle, depending on how you look at it; cats are both alive and dead. This is great for popular culture, which has made "quantum" a buzzword for cool, geek mystification. But it's terrible for those of us who want to understand the world we live in, for there seems to be no easy answer to the simple question, "What is a rock?"

In the first quarter of the twentieth century a theory called *quantum mechanics* was developed to explain quantum physics. This theory has been, ever since its inception, the golden child of science. It is the basis of our understanding of atoms, radiation, and so much else, from the elementary particles and basic forces to the behavior of materials. It also has been, for just as long, a troubled child. From the beginning, its inventors were deeply split over what to make of it. Some expressed shock and misgivings, even outrage. Others declared it a revolutionary new kind of science, which shattered the metaphysical assumptions about nature and our relationship to it that previous generations had thought essential for the success of science.

In these chapters I hope to convince you that the conceptual problems and raging disagreements that have bedeviled quantum mechanics since its inception are unsolved and unsolvable, for the simple reason that the theory is wrong. It is highly successful, but incomplete. Our task—if we are to have simple answers to our simple questions about what rocks are—must be to go beyond quantum mechanics to a description of the world on an atomic scale that makes sense.

This task might seem overwhelmingly difficult, were it not for one almost forgotten and long-ignored aspect of the history of quantum mechanics. Since the very beginning of the quantum era, in the 1920s, there has been an alternative version of quantum physics that does make complete sense. This shadow theory resolves the apparent paradoxes and mysteries of the quantum domain. The scandal—and I believe that term is warranted—is that this alternative form of quantum theory is rarely taught. It is seldom mentioned, either in textbooks for budding physicists or in popularizations for laypeople.

There are several alternative formulations of quantum physics that make consistent sense. The challenge now is to build on these to find the right way to understand quantum physics—the one that nature uses. I believe this will have wide repercussions, because the new form of quantum physics will be the basis of the solutions to many of the outstanding problems in physics. Problems such as quantum gravity and the unification of the forces, on which we have made little definitive progress, are, I believe, foundering because at the foundations of our theorizing is an incorrect theory.

Physicists agree about how the quantum world behaves. We agree that atoms and radiation behave differently than rocks and cats, and we agree that quantum mechanics works to predict some aspects of that behavior. But we don't agree about what it means that our world is a quantum world. It is clear that some radical change in our understanding of nature is required, but we disagree as to what that change needs to be. Some argue that we must give up holding any picture of reality and settle for a theory which describes only the knowledge we can have of the world. Others claim that our notion of reality must be vastly extended to embrace an infinitude of parallel realities.

In fact, neither is necessary. The alternative ways of understanding the quantum world do not require us to give up the idea that physics describes a reality independent of our knowledge of it. Nor do they require that we expand that reality beyond the commonsense notion that there is one world and it is what we see when we look around us. As I'll explain in these pages, commonsense realism, according to which science can aspire to give a complete picture of the natural world as it is, or would be in our absence, is not actually threatened by anything we know about quantum physics.

It is thus both unfortunate and unnecessary that the quantum realm has been presented as mysterious and counterintuitive. One of the aims of this book is to present the alternative quantum theories to laypeople and, by doing so, to lift the mystery and present the quantum world in a way that is intuitive and accessible to people who are not specialists in physics.

I imagine my reader as someone with a strong curiosity about nature, who may follow science through news, blogs, and popular books, but whose education has not included the mathematics usually assumed as the language of physics. Instead I use words and pictures to convey the basic phenomena we find in the quantum world as well as the principles their study has inspired. After an introduction, the book starts with three short chapters which describe the bare-bones basics of quantum physics. These will equip us to explore the diverse conceptual universes which arise from the different forms of quantum theory that have been proposed.

WHAT IS AT STAKE in the argument over quantum mechanics? Why does it matter if our fundamental theory of the natural world is mysterious and paradoxical?

Behind the century-long argument over quantum mechanics is a fundamental disagreement about the nature of reality—a disagreement which, unresolved, escalates into an argument about the nature of science.

Two questions underlie the schism.

First off, does the natural world exist independently of our minds? More precisely, does matter have a stable set of properties in and of itself, without regard to our perceptions and knowledge?

Second, can those properties be comprehended and described by us? Can we understand enough about the laws of nature to explain the history of our universe and predict its future?

The answers we give to these two questions have implications for larger questions about the nature and aim of science, and the role of science in the larger human project. These are, indeed, questions about the boundary between reality and fantasy.

People who answer yes to these two questions are called realists. Einstein was a realist. I am also a realist. We realists believe that there is a real world out there, whose properties in no way depend on our knowledge or perception of it. This is nature—as it would be, and mostly is, in our absence. We also believe that the world may be understood and described precisely enough to explain how any system in the natural world behaves.

If you are a realist, you believe that science is the systematic search for that explanation. This is based on a naive notion of truth. Assertions about objects or systems in nature are true to the extent that they correspond to genuine properties of nature.

If you answer no to one or both of these questions, you are an anti-realist.

Most scientists are realists about everyday objects on the human scale. Things we can see, pick up, and throw around have simple and easily comprehended properties. They exist at each moment somewhere in space. When they move, they follow a trajectory, and that trajectory has, relative to someone describing them, a definite speed. They have mass and weight.

When we tell our partner that the red notebook they are looking for is on the table, we expect that this is simply true or false, absolutely independent of our knowledge or perception.

The description of matter at this level, from the smallest scales we can see with our eyes up to stars and planets, is called classical physics. It was invented by Galileo, Kepler, and Newton. Einstein's theories of relativity are its crowning achievements.

But it is not easy, or obvious, for us to be realists about matter on the scale of individual atoms. This is because of quantum mechanics.

Quantum mechanics is presently our best theory of nature at the atomic scale. That theory has, as I have alluded to, certain very puzzling features. It is widely believed that those features preclude realism. That is, quantum mechanics requires that we say no to one or both of the two questions I asked above. To the extent that quantum mechanics is the correct description of nature, we are forced to give up realism.

Most physicists are not realists about atoms, radiation, and elementary particles. Their belief, for the most part, does not stem from a desire to reject realism on the basis of radical philosophical positions. Instead, it is because they are convinced quantum mechanics is correct and they believe, as they have been taught, that quantum mechanics precludes realism.

If it is true that quantum mechanics requires that we give up realism, then, if you are a realist, you must believe that quantum mechanics is false. It may be temporarily successful, but it cannot be the fully correct description of nature at an atomic scale. This led Einstein to reject quantum mechanics as anything more than a temporary expedient.

Einstein and other realists believe that quantum mechanics gives us an incomplete description of nature, which is missing features necessary for a full understanding of the world. Einstein sometimes imagined that there were "hidden variables" which would complete the description of the world given by quantum theory. He believed that the full description, including those missing features, would be consistent with realism.

Thus, if you are a realist and a physicist, there is one overriding imperative, which is to go beyond quantum mechanics to discover those missing features and use that knowledge to construct a true theory of the atoms. This was Einstein's unfinished mission, and it is mine.

THERE ARE DIFFERENT KINDS of anti-realists, which leads to different views on quantum mechanics.

Some anti-realists believe that the properties we ascribe to atoms and elementary particles are not inherent in those objects, but are created only by our interactions with them, and exist only at the time when we measure them. We can call these radical anti-realists. The most influential of these was Niels Bohr. He was the first to apply quantum theory to the atom, after which he became the leader and mentor to the next generation of quantum revolutionaries. His radical anti-realism colored much of how quantum theory came to be understood.

Another group of anti-realists believes that science, as a whole, does not deal in or talk about what is real in nature, but rather only ever talks about our knowledge of the world. In their view, the properties physics ascribes to an atom are not about that atom; they are instead only about the knowledge we have of the atom. These scientists can be called quantum epistemologists.

And then there are the operationalists, a group of anti-realists who are agnostic about whether there is a fundamental reality independent of us or not. Quantum mechanics, they argue, is not in any case about reality; it is rather a set of procedures for interrogating atoms. It is not about the atoms themselves; it is about what happens when atoms come into contact with the big devices we use to measure them. Heisenberg, the best of Bohr's protégés, who invented the equations of quantum mechanics, was, at least partly, an operationalist.

In contrast to the disputes between radical anti-realists, quantum epistemologists, and operationalists, all realists share a similar perspective—we agree about the answer to both questions I posed above. But we differ on how we answer a third question: Does the natural world consist mainly of the kinds of objects that we see when we look around ourselves, and the things that constitute them? In other words, is what we see when we look around typical of the universe as a whole?

Those of us who say yes to this question can call ourselves simple or naive realists. I should alert the reader that I use the adjective "naive" to mean strong, fresh, uncomplicated. For me, a view is naive if it is not in need of sophisticated arguments or convoluted justifications. I would argue that a naive realism is, whenever possible, to be preferred.

There are realists who are not naive in this sense. They believe that reality is vastly different from the world we perceive and measure.

An example of such a view is the Many Worlds Interpretation, which teaches that the world we perceive is only one of a vast and ever-growing number of parallel worlds. Its proponents call themselves realists, and they have some claim to that designation by virtue of their answering yes to the first two questions. But, in my opinion, they are realists only in the most technical, academic sense. They may perhaps be called magical realists, for they believe that what is real is far beyond the world we perceive. Magical realism in this sense is almost a form of mysticism, for it implies that the true world is hidden from our perception.

Is it possible to formulate a theory of atoms that is realist in the most general and naive sense, and so answers yes to all three questions? It is, and that is the story I want to tell in this book. But that theory is not quantum mechanics, and if it is right, then quantum mechanics is wrong, in the sense that quantum mechanics must then give a very incomplete description of nature.

Part of the story I want to tell here is how this naively realistic theory of nature was pushed aside, while a theory that required us to embrace either anti-realism or mysticism thrived. But I will end on a hopeful note, by sketching a way we may progress to a realist view of nature that encompasses the quantum.

THIS ALL MATTERS because science is under attack in the early twenty-first century. Science is under attack, and with it the belief in a real world in which facts are either true or false. Quite literally, parts of our society appear to be losing their grip on the boundary between reality and fantasy.

Science is under attack from those who find its conclusions inconvenient for their political and business objectives. Climate change should not be a political issue; it is not a matter of ideology, but an issue of national security, and should be treated as such. It is a real problem, which will require evidence-based solutions. Science is also under attack from religious fundamentalists who insist ancient texts are the teachings of unchanging truths by God.

In my view, there is little reason for conflict between most religions and science. Many religions accept—and even celebrate—science as the way to knowledge about the natural world. Beyond that, there is mystery enough in the existence and meaning of the world, which both science and religion can inspire us to discuss, but neither can resolve.

All that is required is that religions not attack or seek to undermine those scientific discoveries which are considered to be established knowledge because they are supported by overwhelming evidence, as judged by those educated sufficiently to evaluate their validity. This is indeed the view of many religious leaders from all faiths. In return, scientists should view these enlightened leaders as allies in the work for a better world.

In addition, science is under attack from a fashion among some humanist academics—who should know better—who hold that science is no more than a social construction that yields only one of an array of equally valid perspectives.

For science to respond clearly and strongly to these challenges, it must itself be uncorrupted by its own practitioners' mystical yearnings and metaphysical agendas. Individual scientists may be—and, let's face it, sometimes are—motivated by mystical feelings and metaphysical preconceptions. This doesn't hurt science as long as the narrow criteria that distinguish hypothesis and hunch from established truth are universally understood and adhered to.

But when fundamental physics itself gets hijacked by an anti-realist philosophy, we are in danger. We risk giving up on the centuries-old project of realism, which is nothing less than the continual adjustment, bit by bit as knowledge progresses, of the boundary between our knowledge of reality and the realm of fantasy.

One danger of anti-realism is to the practice of physics itself. Anti-realism lowers our ambition for a totally clear understanding of nature, and hence weakens our standards as to what constitutes an understanding of a physical system.

In the wake of the triumph of anti-realism about the atomic world, we have had to contend with anti-realist speculations about nature on the largest possible scale. A vocal minority of cosmologists proclaims that the universe we see around us is only a bubble in a vast ocean called the multiverse that contains an infinity of other bubbles. And, whereas it is safe to hypothesize that the galaxies we can see are typical of the rest of our universe, one must regard the other invisible bubbles as governed by diverse and randomly assigned laws, so our universe is far from typical of the whole. This, together with the fact that all, or almost all, of the other bubbles are forever out of range of our observations, means the multiverse hypothesis can never be tested or falsified. This puts this fantasy outside the bounds of science. Nonetheless, this idea is championed by not a few highly regarded physicists and mathematicians.

It would be a mistake to confuse this multiverse fantasy for the Many Worlds Interpretation of quantum mechanics. They are distinct ideas. Nonetheless, they share a magical-realist subversion of the aim of science to explain the world we see around us in terms of only itself. I would suggest that the harm done to clarity about the aim and purpose of science by the enthusiastic proponents of the multiverse would not have been possible had not the majority of physicists uncritically adopted anti-realist versions of quantum physics.

Certainly, quantum mechanics explains many aspects of nature, and it does so with supreme elegance. Physicists have developed a very powerful tool kit for explaining diverse phenomena in terms of quantum mechanics, so when you master quantum mechanics you control a lot about nature. At the same time, physicists are always dancing around the gaping holes that quantum mechanics leaves in our understanding of nature. The theory fails to provide a picture of what is going on in individual processes, and it often fails to explain why an experiment turns out one way rather than another.

These gaps and failures matter because they underlie the fact that we have gotten only partway toward solving the central problems in science before seeming to run out of steam. I believe that we have not yet succeeded in unifying quantum theory with gravity and spacetime (which is what we mean by quantizing gravity), or in unifying the interactions, because we have been working with an incomplete and incorrect quantum theory.

But I suspect that the implications of building science on incorrect foundations go further and deeper. The trust in science as a method to resolve disagreements and locate truth is undermined when a radical strand of anti-realism flourishes at the foundations of science. When those who set the standard for what constitutes explanation are seduced by a virulent mysticism, the resulting confusion is felt throughout the culture.

I WAS PRIVILEGED to meet a few of the second generation of the founders of twentieth-century physics. One of the most contradictory was John Archibald Wheeler. A nuclear theorist and a mystic, he transmitted the legacies of Albert Einstein and Niels Bohr to my generation through the stories he told us of his friendships with them. Wheeler was a committed cold warrior who worked on the hydrogen bomb even as he pioneered the study of quantum universes and black holes. He was also a great mentor who counted among his students Richard Feynman, Hugh Everett, and several of the pioneers of quantum gravity. And he might have been my mentor, had I had better judgment.

A true student of Bohr, Wheeler spoke in riddles and paradoxes. His blackboard was unlike any I'd ever encountered. It had no equations, and only a few elegantly written aphorisms, each set out in a box, distilling a lifetime of seeking the reason why our world is a quantum universe. A typical example was "It from bit." (Yes, read it again—slowly! Wheeler was an early adopter of the current fashion to regard the world as constituted of information, so that information is more fundamental than what it describes. This is a form of anti-realism we will discuss later.) Here is another: "No phenomenon is a real phenomenon until it is an observed phenomenon." Here is the kind of conversation one had with Wheeler: He asked me, "Suppose when you die and go up before Saint Peter for your final, final exam, he asks you just one question: 'Why the quantum?'" (I.e., why do we live in a world described by quantum mechanics?) "What will you say to him?"

Much of my life has been spent searching for a satisfying answer to that question. As I write these pages, I find myself vividly recalling my first encounters with quantum physics. When I was a seventeen-year-old high school dropout, I used to browse the shelves at the University of Cincinnati Physics Library. There I came upon a book with a chapter by Louis de Broglie (we will meet him in chapter 7), who was the first to propose that electrons are waves as well as particles. That chapter introduced his pilot wave theory, which was the first realist formulation of quantum mechanics. It was in French, a language I read fitfully after two years of high school study, but I recall well my excitement as I understood the basics. I still can close my eyes and see a page of the book, displaying the equation that relates wavelength to momentum.

My first actual course in quantum mechanics was the next spring at Hampshire College. That course, taught by Herbert Bernstein, ended with a presentation of the fundamental theorem of John Bell,[1] which, in brief, demonstrates that the quantum world fits uneasily into space. I vividly recall that when I understood the proof of the theorem, I went outside in the warm afternoon and sat on the steps of the college library, stunned. I pulled out a notebook and immediately wrote a poem to a girl I had a crush on, in which I told her that each time we touched there were electrons in our hands which from then on would be entangled with each other. I no longer recall who she was or what she made of my poem, or if I even showed it to her. But my obsession with penetrating the mystery of nonlocal entanglement, which began that day, has never left me; nor has my urgency to make better sense of the quantum diminished over the decades since. In my career, the puzzles of quantum physics have been the central mystery to which I've returned again and again. I hope in these pages to inspire in you a similar fascination.

The story I tell in this book is shaped like a play in three acts. Part 1 teaches the basic concepts of quantum mechanics we will need while tracing the story of its invention. The main theme here is the triumph of the anti-realists, led by Bohr and Heisenberg, over the realists, whose champion was Einstein. Please note that the story I tell here is just a sketch; the real history is far more complex. Part 2 traces the revival of realist approaches to quantum mechanics, beginning in the 1950s, and explains their strong and weak points. The heroes here are an American physicist named David Bohm and an Irish theorist, John Bell.

The conclusion of part 2 will be that realist approaches are possible, and work well enough to undermine the claims that quantum physics requires us all to become anti-realists. Still, for me, none of these approaches have the ring of truth. I believe we can do better; indeed, for reasons I will explain, I would venture that the correct completion of quantum mechanics will also solve the problem of quantum gravity, as well as give us a good cosmological theory. Part 3 introduces contemporary efforts to construct this realist theory of everything, some mine, some by others.

Welcome to the quantum world. Feel at home, for it is our world, and it is our good fortune that its mysteries remain for us to solve.

PART 1

AN ORTHODOXY OF THE UNREAL

ONE

Nature Loves to Hide

Reality is the business of physics.

—ALBERT EINSTEIN

Quantum mechanics has been the core of our understanding of nature for nine decades. It is ubiquitous, but it is also deeply mysterious. Little of modern science would make sense without it. But experts have a hard time agreeing what it asserts about nature.

Quantum mechanics explains why there are atoms, and why those atoms are stable and have distinct chemical properties. Quantum mechanics also explains how atoms combine into diverse molecules. As a result, it is the basis for how we understand the shapes and interactions of those molecules. Life would be incomprehensible without the quantum. From the behavior of water to the shapes of proteins to the fidelity and transmittal of information by DNA and RNA, everything in biology depends on the quantum.

Quantum mechanics explains the properties of materials, such as what makes a metal a conductor of electricity, while another is an insulator. It explains light and radioactivity, and is the basis of nuclear physics. Without it we wouldn't understand how the stars shine. Nor could we have invented the chips or the lasers on which so much of our technology is based. Quantum mechanics is the language that we use to write the standard model of particle physics, which contains all we know about the elementary particles and the fundamental forces by which they interact.

According to our best theory of the early universe, all matter, along with the patterns that eventually coalesced into the galaxies, was yanked into existence from the quantum randomness of the vacuum of empty space by the rapid expansion of the universe. I don't expect the reader to understand precisely what this means, but perhaps the words evoke an image. In any case, if this is right, then without quantum physics there would literally be nothing except empty spacetime.

Yet for all its success, there is a stubborn puzzle at the heart of quantum mechanics. The quantum world behaves in ways that challenge our intuition. It is often said that in quantum physics an atom can be in two places at once, but that is only the start; the full story is far weirder than that. If an atom can be here or there, we must speak of states in which it is, somehow, simultaneously both here and there. This is called a *superposition*.

If you are new to the quantum world, you are undoubtedly wondering what it means for an atom to be somehow both here and there. Don't be discouraged if you find this confusing. You are absolutely right to wonder what it means. This is one of the central mysteries of quantum mechanics. It is enough, for now, if you just accept this as a mystery, to which we attach the term "superposition." Later we will be able to demystify it.

Here is a first step. When we say that a quantum particle is in a "superposition of being here and there," this is related to the wavelike nature of matter, for a wave is a disturbance that is spread out, and so it can be both here and there.

We speak of elementary particles, but everything quantum, including atoms and molecules, is both a particle and a wave. Here is a taste of what that means. If we do an experiment that asks where an atom is, the result will be that it is somewhere definite. But between measurements, when we are not looking for it, it turns out to be impossible to project where it might be. It is as if the likelihood or propensity of finding the particle spreads as a wave when we are not looking. But as soon as we look again, it is always somewhere.

Imagine playing a game of hide-and-seek with an atom. We open our eyes, or turn on a detector, and we see it somewhere. But when we close our eyes it dissolves into a wave of potentiality. Open our eyes again and it is always somewhere.

Another feature unique to the quantum world is called *entanglement*. If two particles interact, and then move apart, they remain intertwined in the sense that they seem to share properties which cannot be broken down to properties each enjoys individually.

We can stretch our imagination to apply these new concepts to atoms and molecules which are too small to see directly. We must study them indirectly, and to do that we employ large and complex measurement devices.

Those measurement devices are part of the everyday, familiar world of large objects. One thing we can be sure of is that big everyday things display none of the bizarre behavior quantum mechanics describes. A chair is here or it is there, never in a combination of such states. When we wake up in the middle of the night in a strange hotel room, we may be unsure where the chair is, but we can be sure it is somewhere. And after we collide with it in the dark, our future does not become entangled with its future.

In the world as we experience it, cats are either alive or dead, even if they are locked in a box. When we open the box, the cat does not suddenly resolve from a combination of dead and alive to dead. If we find it dead it will likely have been so for some time, as we will instantly smell.

Ordinary objects appear to share none of the quantum weirdnesses of the atoms of which they are made. This seems obvious, but it raises a mystery. Quantum mechanics is the core theory of nature. As such it must be universal. If it applies to an atom it must apply to two atoms, or ten or ninety. And we have excellent experimental evidence that it does. Delicate experiments, in which large molecules are put in quantum superpositions, show us that they are just as quantum weird as electrons. For one thing, they diffract and interfere as waves.

But then quantum mechanics must apply to the vast collections of atoms that make up you or me or our cat or the chair on which she is perched. But it doesn't seem to. Nor does quantum mechanics appear to apply to any of the instruments and machines we employ to image the atoms and reveal their quantum weirdnesses. How can this be?

In particular, when we measure a property of an atom, we employ big devices. The atoms may be in superpositions of states and so be several places at once, but the measuring instrument always indicates just one out of the possible answers to the questions we pose. Why is that? Why does quantum mechanics not apply to the very devices we use to measure quantum systems?

This is called the *measurement problem*. It has been controversial and unresolved since the 1920s. The fact that, after all this time, we have found no agreement among experts means there is something basic about nature we have yet to understand.

So there is somewhere a transition between the quantum world, in which an atom can be several places at once, and the ordinary world, in which everything is always somewhere. If a molecule made from ten or ninety atoms can be described by quantum mechanics, but a cat cannot, then somewhere between the two there is a line delineating where the quantum world stops. An answer to the measurement problem would tell us where that line is and explain the transition.

There are people who are sure they know the answer to the measurement problem. We will meet some of them and their ideas later on. We will want to look out for what price we have to pay to expunge this quantum insanity from our understanding of the world.

BROADLY SPEAKING, the people who aim to address the mysteries of quantum mechanics fall into two classes.

The first group assumes that the theory as it was formulated in the 1920s is essentially correct. They believe the problem is not with quantum theory; it is instead with how we understand it or speak about it. This strategy to mitigate the strangeness of quantum mechanics goes back to some of the founders, beginning with Niels Bohr.

Niels Bohr was a Danish physicist who, while still in his twenties, was the first to apply quantum theory to atoms. As he grew older he became the de facto leader of the quantum revolution, partly due to the attractiveness of his ideas and partly because he educated and mentored many of the young quantum revolutionaries.

The second group has concluded that the theory is incomplete. It can't be made sense of because it is not the whole story. They seek a completion of the theory that will tell us the rest of the story and, by doing so, resolve the mysteries of quantum mechanics. This strategy goes back to Albert Einstein.

More than anyone else, Einstein was responsible for initiating the quantum revolution. He was the first to articulate the dual nature of light as a particle and a wave. He is by now better known for his theory of relativity, but his Nobel Prize was for his work on quantum theory, and he himself admitted that he spent much more time on quantum theory than on relativity. Yet, even if he initiated the quantum revolution, Einstein did not become one of its leaders, because his realism required that he reject the theory as it was developed in the late 1920s.

In the language introduced in the preface, those in the first group are mostly anti-realists or magical realists. Realists find themselves in the second group.

Those who argue for the incompleteness of quantum mechanics point to the fact that in most cases it can only make statistical predictions for the results of experiments. Rather than saying what will happen, it gives probabilities for what might happen. In a letter to his friend Max Born in 1926, Einstein wrote:

> Quantum mechanics is certainly imposing. But an inner voice tells me that it is not yet the real thing. The theory says a lot, but does not really bring us any closer to the secret of the 'old one'. I, at any rate, am convinced that *He* is not playing at dice.[1]

Einstein was also friends with Niels Bohr, and their divergent responses to quantum mechanics fueled a passionate debate between them that lasted more than forty years, till Einstein's death. It continues between their intellectual descendants to this day. Einstein was the first person to clearly articulate the need for a revolutionary new theory of atoms and radiation, but he was unable to accept that quantum mechanics was that theory. His first response to quantum mechanics was to argue that it was inconsistent. When that failed, he argued instead that quantum mechanics gives an incomplete description of nature, which leaves out an essential part of the picture.

I believe that Einstein was unable to accept quantum mechanics as a definitive theory because he had exceedingly high aspirations for science. He was driven by the hope of transcending subjective opinion and discovering a true mirror of nature that exhibits the essence of reality in a few timeless mathematical laws. For him, science aimed to capture the true essence of the world, and that essence is independent of us and can have nothing to do with what we believe or know about it.

Einstein, of all people, must have felt he had the right to demand this because he had achieved it in his discoveries of special and general relativity. Having laid the groundwork for quantum physics, he sought to capture the essence of the atomic world in a complete description of atoms, electrons, and light.

Bohr replied that atomic physics required a revolutionary revision in how we understand what science is, as well as in our conception of the relationship between reality and our knowledge of it. This stemmed from the fact that we are a part of the world, so we must interact with the atoms we seek to describe.

Bohr asserted that once we absorbed this revolutionary change in our thinking, the completeness of quantum mechanics would be unavoidable, because it was built into our being participants in the world we seek to describe. From Bohr's perspective, quantum theory is complete in the sense that there is no more-complete description of the world to be had.

If we refuse these philosophical revolutions and insist on

maintaining an old-fashioned, commonsense view of reality and its relation to our observations and knowledge, we have to pay a different kind of price. We have to contemplate that we are wrong about some aspect of nature. We have to find out which common assumption is wrong and replace it with a radically new physical hypothesis that opens the way to a new theory that will complete quantum mechanics.

Thanks to a combination of theory and experiment, starting with a paper by Einstein and two collaborators in 1935, we know one aspect of this completion. The new theory must violate the commonplace assumption that things interact only with other things that are near them in space.

This assumption is called locality. A big part of the story I will be telling in later chapters is how this commonsense idea must be transcended in the theory which will replace quantum mechanics.

THIS BOOK HAS THREE PURPOSES. First, I want to explain to laypeople just what the puzzles at the heart of quantum mechanics are. After more than a century of studying quantum physics, it is remarkable that there continues to this day to be no agreement on the solution of these puzzles.

But having explained the reasons for the debate in a way that is fair to both sides, I will not stay impartial. In the great debate about whether quantum mechanics is the last word or not, I side with Einstein. I believe that there is a layer of reality deeper than that described by Bohr, which can be understood without compromising old-fashioned notions of reality and our ability to comprehend and describe it.

Thus, my second purpose is to advocate a point of view about the puzzles of quantum mechanics. This is that the problems can be resolved only by progress in science which will uncover a world beyond quantum mechanics. Where quantum mechanics is mysterious and confusing, this deeper theory will be entirely comprehensible.

I can make this claim because we have known since the invention of quantum mechanics how to present the theory in a way that dissolves the mysteries and resolves the puzzles. In this approach, there is no challenge to our usual beliefs in an objective reality, a reality unaffected by what we know or do about it, and about which it is possible to have complete knowledge. In this reality, there is just one universe, and when we observe something about it, it is because it is true. This can justly be called a realist approach to the quantum world.

An anti-realist approach ascribes the mysteries of quantum mechanics to subtleties having to do with how we gain knowledge about nature. Such approaches have radical proposals to make about *epistemology*, which is the branch of philosophy concerned with how we know things. Realist approaches assume we are able to arrive sooner or later at a true representation of the world and so are deliberately naive about epistemology. Instead, realists are interested in *ontology*, which is the study of what exists. By contrast, anti-realists believe we cannot know what really exists, apart from our representation of the knowledge we have of the world, which is gained only through interacting with it.

So I will endeavor to reassure readers that quantum mechanics can be understood completely within a realist perspective in which the external world can be completely comprehended as independent from us. There is no mysterious effect of the observer on the observed. Reality is out there, recalcitrant to our will and the choices we make. That reality is fully comprehensible. And that reality consists of a single world.

The existence of these realist approaches to quantum mechanics does not by itself mean that the philosophically more extravagant proposals are wrong. But it does mean that there is no strong scientific reason to believe in them, because realism is always to be preferred in science, when it can be achieved.

Why, then, is so much of the talk about quantum theory inspired by the weirder ideas in which reality depends on our knowledge of it or there are multiple realities? This is a problem for historians of ideas. One such historian, Paul Forman, has tied the dominance of Bohr and Heisenberg's anti-realist philosophy within the scientific community in the 1920s and 1930s to the embrace of chaos and irrationality advocated by Spengler and others in the wake of the First World War.

That history is fascinating, but it is for scholars to do justice to it. I am not a scholar, I am a scientist, and this brings me to my third purpose in writing this book.

I have been on Einstein's side in the search for a deeper but simpler reality behind quantum mechanics since first reading him on the subject as a high school dropout. My journey in physics began with reading Einstein's autobiographical notes, where, in the last few years of his life, in the 1950s, he reflected on the two main tasks he felt were left incomplete in physics. These were to make sense of quantum physics and, after that, to unify the new understanding of the quantum with gravity, by which he meant his general theory of relativity. I recall thinking that maybe I could try to help. I was unlikely to succeed, but perhaps here was something worth striving for.

After, as it were, getting my mission from reading Einstein's autobiographical notes, I found that book by de Broglie, talked my way into a good college, found great teachers, and got lucky several times in my applications for graduate school and beyond. I'm having a wonderful life, and as a scientist on the frontier, I've had many chances to take a shot on goal, aimed at solving Einstein's two big questions.

I haven't succeeded, at least so far. Very unfortunately, neither has anyone else. At the same time, over the past several decades there has at least been progress toward understanding the problem. That is not nearly as good as it would be to solve the problem, but neither is it nothing. We know much better than Einstein did the obstacles that a theory that transcends the limits of quantum mechanics must overcome. And because of that, some very interesting proposals and hypotheses have been put forward, which may frame the deeper theory for which we search.*

I have been thinking about the question of how to go beyond quantum mechanics since the mid-1970s, and I've never been more excited and optimistic about the prospects for success. So this is my third reason for writing this book, which is to bring to a wider audience a report from the front in our search for the world beyond the quantum.

* A note to my expert readers: Quantum foundations is presently a very lively subject, with many exciting developments both experimental and theoretical. Many proposals compete to resolve the puzzles we will meet here. I should warn the reader that our path through these frontiers will be a narrow one, and there are many exciting ideas and results that I do not mention here. Had I tried to review the whole field, or include all the latest supremely clever advances, the result would have been a less accessible book. My first aim is to introduce the world of quantum phenomena, not the full spectrum of competing interpretations of those phenomena. I apologize in advance to those experts who don't find their preferred version of quantum physics here, and encourage them to write their own books. I also apologize to the historians. I am not a scholar, and the stories I tell are creation myths, handed down from teacher to student, originating, in some cases, with the founders themselves.

TWO

Quanta

If we break quantum mechanics down to its most essential principle, it is this:

We can only know half of what we would need to know if we wanted to completely control, or precisely predict, the future.

This disrupts the basic ambition of physics, which is to be able to predict the future. It was hoped that this power would follow if only we could give the physical world a complete description. By describing fully the motion of every particle and the action of every force, we would be able to work out exactly what would happen in the future. Before quantum mechanics was formulated in the 1920s, we physicists were confident that if we could learn the laws that govern the fundamental particles, we would be able to predict and explain everything that happened in the world.

The hypothesis that the future is completely determined by the laws of physics acting on the present configuration of the world is called determinism. This is an extraordinarily powerful idea, whose influence can be seen in diverse fields. If you appreciate the extent to which determinism dominated thought in the nineteenth century, you can begin to understand the revolutionary impact of quantum mechanics across all fields, because quantum mechanics precludes determinism.

To emphasize this point, I like to quote from Tom Stoppard's play *Arcadia*, in which his precocious heroine, Thomasina, explains to her tutor:

> If you could stop every atom in its position and direction, and if your mind could comprehend all the actions thus suspended, then if you were really, *really* good at algebra you could write the formula for all the future; and although nobody can be so clever as to do it, the formula must exist just as if one could.[1]

A complete description of nature, at a given time, is called a *state*. For example, if we think of the world as composed of particles whizzing around, the state tells us where each of them is, and how fast and in what direction each is moving, at that moment.

The power of physics comes from its laws, which dictate how nature changes in time. They do this by transforming the state of the world as it is now to the state at any future time. A law of physics functions in some ways like a computer program: it reads in input and puts out output. The input is the state at a given time; the output is the state at some future time.*

Along with the computation comes an explanation of how the

* The metaphor of the universe as a computer is helpful for illustrating determinism, but is on the whole misleading, as I will argue below.

world changes in time. The law acting on the present state *causes* the future states. A successful prediction of the future state is taken as a validation of that explanation. The prediction is deterministic, in that a precise input leads to a precise output. This confirms a belief that the information that went into describing the state is in fact a complete description of the world at one moment of time.

This concept of a law is basic to a realist conception of nature and, as such, transcends any one theory. Newtonian mechanics and Einstein's two theories of relativity all work the same way. One applies the law to the state at an initial time, and it transforms that state to the state at some future time. This schema for explaining nature was invented by Newton, so we call it the Newtonian paradigm.

It is also worth mentioning that in almost all cases so far known, the laws are reversible. One can input the state at some future time and run the law backward to output the state at an earlier time. (The issue of the reversibility of time and of the fundamental laws is a central concern of chapters 14 and 15.)

It is often the case that the information needed to completely describe the state of a physical system comes in pairs. Position and momentum.* Volume and pressure. Electric field and magnetic field. We need both to predict the future. Quantum mechanics says we can know only one.

This means we can't precisely predict the future. That is just the first of the blows to our comfortable intuitions that we will have to absorb from quantum theory.

Which member of each pair is the one that can be known? Quantum mechanics says you choose! This is the basis of its challenge to realism.

* Momentum will be defined shortly, but roughly, a body's momentum is proportional to both its speed and mass.

There is more to say about the impossibility of predicting the future. To get there, let's take advantage of the great generality quantum mechanics claims, and speak a bit abstractly. We want to describe some physical system in terms of a pair of variables—we will call them A and B. Quantum physics asserts a two-part principle.

1. If we knew both A and B at a given time, we could precisely predict the future of the system.
2. We can choose to measure A or we can choose to measure B; and in each case we will succeed. But we can't do better. We cannot choose to simultaneously measure both A and B.

As I have stated it, this is a prohibition of what we can measure; but, if we prefer, we can express it as a prohibition of what we can know about the system.

But wait, why can't you measure A and then, at a later time, measure B? You can. But your measurement of B will render irrelevant (for the purpose of predicting the future) your past knowledge of A. One way this can happen is that after the measurement of B, the value of A is randomized. We cannot measure B without disrupting the value of A, and vice versa. Thus, if we measure A, then B, then A again, the value of A we get the second time will be random, and hence unrelated to the value we got the first time we measured A.

1. and 2. together are called the *principle of non-commutativity*. Two actions are said to *commute* if it doesn't matter in what order we do them. If the action that is done first matters, we say they are non-commutative. It doesn't matter (except to a few fanatics) in what order you put milk and sugar into coffee; they commute.

Getting dressed involves non-commutative operations; the order in which you put on your underwear and pants matters. But it doesn't matter which sock you put on first, or whether you put your socks on first, partway through the process, or last. So putting on socks commutes with everything except putting on shoes. (The mathematically minded will understand this as an application of algebra to topology.)

What if we allow there to be some specified amount of uncertainty in the measurement of A? Then we can measure B, but only up to some accuracy. These uncertainties are reciprocal—the better we know A, the worse we can know B, and vice versa.

For example, let's suppose that A is the position of a particle. Then B is its momentum. Suppose we do a measurement that tells its location to within a meter. Then we can measure the momentum to a corresponding uncertainty. If we increase the uncertainty in A, then we can make the measurement of B more precise and vice versa. This gives us a principle called, not surprisingly, the *uncertainty principle*.*

$$(\text{Uncertainty in A}) \times (\text{Uncertainty in B}) > \text{a constant}$$

Applied to position and momentum, it reads

$$(\text{Uncertainty in position}) \times (\text{Uncertainty in momentum}) > \text{a constant}$$

Physics is like a college campus where every building is named after someone. The constant is named after Max Planck, and the uncertainty principle is named after Werner Heisenberg.

* > means "is larger than."

The uncertainty principle is quite powerful, as is shown by this important consequence. Let's go back to the scenario in which you measure A, then you measure B, then you measure A again. As I said above, once you know the result of measuring B, the second measurement of A is randomized; it is no longer equal to the original value of A. But suppose that, just before you remeasure A, you do something to forget what the value of B was. Then the system remembers—yes, that word is the one we use to describe this situation—the original value of A.

This is called *interference*. It is allowed by the uncertainty principle because once you forget the measurement of B, B's uncertainty is very large, so A's uncertainty can be small.

But how can we undo a measurement? Let me give a fanciful example. There are many simple cases in which A and B each have two possible values. Let the systems we study be people, and let A be political identity, which we will simplify to be a binary choice: either left-wing or right. I will let B be pet preference, cat lovers versus dog lovers. We now play a game in which a person can't have both a definite pet preference and a political identity. We go to a party where everyone has left-wing views and ask each whether they are a cat person or a dog person. We put the cat lovers in the living room and the dog lovers in the kitchen. If we go into either room and inquire about their political views, then half will now be right-wingers. That is what must happen if political identity and pet preference don't commute.

But let's afterward call everyone together into the dining room. We let them mingle for a while, then we go in and pick a random person. They could have come from either the living room or the kitchen, we don't know which, so we've lost track of their pet preference. Then, when we ask them about politics, we find they are all left-wingers again.

These principles are entirely general. A and B are often the answers to yes/no questions. But in the original case, A was the position of an elementary particle, say an electron, and B was the momentum of the particle.

Momentum is one of those words that functions as a barrier to comprehension, so let's take a moment to define it.

In physics we often have to refer to the speed and the direction of motion of a particle. We combine these into one quantity which we call the *velocity*. You can think of a particle's velocity as an arrow that points in the direction of its motion. The faster the speed, the longer the arrow.

To survive a collision you want to experience as little force as possible. The force a truck will impart on a car is proportional to the truck's change of speed. But it's also proportional to the mass of the truck. You'd rather collide with a Ping-Pong ball than a truck, even if they are traveling toward you at the same speed. To express this, physicists define momentum as the product of the mass times the velocity. This is also an arrow pointing in the direction of motion, only now the length is proportional to both the speed and the mass.

Momentum is a central concept in physics because it is conserved. That means that in any processes at all, we can add up the momenta of the various particles involved at the beginning, and, no matter what happens, the resulting total momentum won't change in time. Before, during, and after a collision, the total momentum will be the same. What happens in a collision is that momentum is exchanged from one body to another. This change of momentum is experienced as a force.

Energy is another conserved quantity. The total energy of a system of particles never changes in time. When particles interact, one may gain energy while the rest lose energy. But the total energy remains the same; none is created or destroyed.

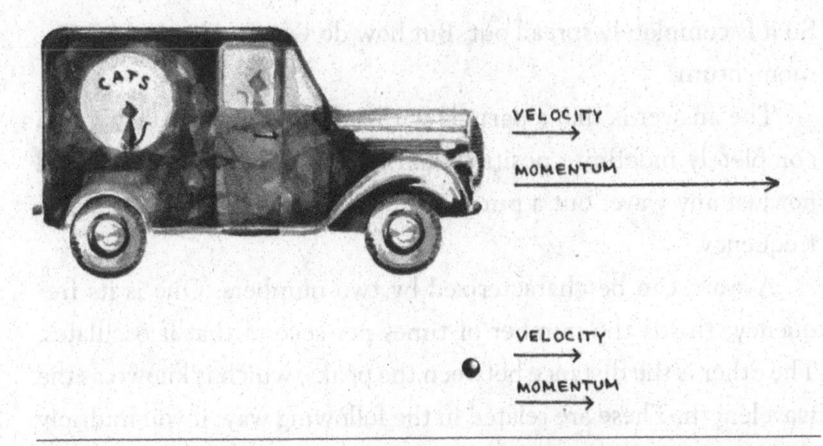

FIGURE 1. A truck carries much more momentum than a Ping-Pong ball going the same velocity, because its mass is so much greater, and the momentum is the product of the mass and the velocity.

Energy and momentum are related. We won't need the exact relation, but we need to know that a particle that is moving freely, and has an exact value of momentum, also has an exact energy.

The uncertainty principle then says that we can't know both the position and momentum of a body at the same time. This means we can't make a precise prediction of its future, because to do so we would need to know both where something is and how fast and in what direction it is moving, with complete accuracy.

If we want to develop an intuition about how quantum particles behave, we will need to be able to visualize a particle with a definite position, but, because of the uncertainty principle, no definite momentum or velocity. This is not hard: visualize the particle being somewhere momentarily. In the next moment it will also be somewhere definite, just somewhere else. Because its momentum is indefinite, it jumps around randomly.

But how do we visualize a particle with a definite momentum, but a completely indefinite position? This seems more challenging. If you look for it, you have an equal chance of finding it anywhere.

So it is completely spread out. But how do we visualize its definite momentum?

The answer is that a particle with a definite momentum, but a completely indefinite position, can be visualized as a wave. And not just any wave, but a pure wave, one which vibrates at a single frequency.

A wave can be characterized by two numbers. One is its frequency; this is the number of times per second that it oscillates. The other is the distance between the peaks, which is known as the wavelength. These are related in the following way: if you multiply these two numbers together, you get the speed at which the wave is traveling. Thus a wave which oscillates with a single frequency will also have a definite wavelength.

Quantum mechanics asserts that the momentum of the particle and the wavelength of the wave that represents it are related in a simple way, which is that they are inversely proportional. That is,

$$\text{wavelength} = h/\text{momentum}$$

h is the same Planck's constant that came into the uncertainty relations.

Let us assume for a moment that no force acts on our particle, perhaps because it is very far from everything else. In the absence of forces, a particle with a definite momentum also has a definite energy. That energy is in turn related to the frequency of the wave, in that they vary proportionately.

$$\text{Energy} = h \times \text{frequency}$$

These relations and correspondences are universal. Everything in the quantum world can be viewed as both a wave and a particle.

This is a direct consequence of the basic principle that we can measure the particle's position or measure its momentum, but we cannot measure both at the same time.

When we wish to measure its position, we visualize it as a particle, localized, but just momentarily, at a point in space. The momentum is completely uncertain, so the next moment, if we look again, we will find it has randomly jumped somewhere else. It can't remain in one place because, if it did, it would have a definite value of momentum, namely zero.

If, on the other hand, we choose to measure the particle's momentum, we will discover it has some definite value. It is nowhere in particular, so we visualize it as a wave, but one with a definite wavelength and frequency, according to the relations just mentioned.

What is so crazily fabulous about this is that waves and particles are quite different. A particle always has a definite position, localized somewhere in space. Its motion traces out a path through space, what we call its trajectory. Moreover, according to Newtonian physics, at each moment a particle also has a definite velocity and, consequently, a definite momentum. A wave is almost the opposite. It is delocalized; it spreads out as it travels, occupying all the space available to it.

But now we are learning that waves and particles are different sides of a duality, that is, different ways of visualizing one reality. A single reality with a dual nature: a duality of waves and particles.

A quantum particle can have a position. We ask where it is, and we will find it somewhere. But a quantum particle never has a trajectory, because, if we know where it is, where it will be next is completely uncertain. We must get used to thinking of particles at definite positions which are not points on trajectories. Similarly, if we measure a momentum we will always find a value. But then it's a wave, spread out everywhere. Where we will find the particle, if we next measure its position, is completely uncertain.

This scheme, it must be admitted, has an incredible elegance. But what is most compelling is its universality. It applies to light, it applies to electrons, and it applies to all the other elementary particles known. It applies to combinations of those particles, such as atoms and molecules. It has worked successfully to describe the motions of large molecules, such as buckyballs and proteins. There is no case of an experiment that was sensitive enough to reveal the quantum nature of an object, but failed to do so. At least so far, size and complexity provide no limit. We do not yet know if the wave-particle duality applies to people or cats or planets or stars, but there is no reason known why it definitely can't.

In all these cases the effect is the same: we can only know half of what we would need to know to precisely predict the future.

THREE

How Quanta Change

In the first lecture of his course on quantum mechanics, my teacher Herbert Bernstein asserted that physics is the science of everything. Our goal in physics is to find the most general laws of nature, from which the multitude of phenomena exhibited by nature may all be explained.

Quantum mechanics explains the widest variety of phenomena of any theory so far. At the same time, it greatly restricts the questions that can be asked of any particular phenomenon. We have already encountered one kind of limitation: that we can know only half what we would need to know about a system to make precise predictions for its future. As a result, we must give up describing exactly what goes on in individual atoms in favor of statistical predictions, which apply only to averages taken over many cases. Hence, to believe in quantum theory we must give up the ambition to precisely predict the future.

Most physicists have given up those ambitions in the face of the success of quantum mechanics. But I believe that this is short-sighted and there is a deeper level of reality to be discovered, the mastery of which will restore our ambitions for a complete understanding of nature.

Another restriction limits the range of quantum theory. We can express this in a principle I call the *subsystem principle*:

> ***Any system quantum mechanics applies to must be a subsystem of a larger system.***

One reason for this is that quantum mechanics refers only to physical quantities which are measured by measuring instruments, and these must be outside the system being studied. Further, the results of these measurements are perceived and recorded by observers, who are also not part of the system being studied.

Most of us approach science with the naive expectation that it will tell us what is real. We can follow John Bell and call a real property of a system a *beable*: it is part of what is. Bell coined the word as a contrast to the term *observables,* which is what anti-realists want out of a theory.

"Observables" and "beables" are loaded terms, whose use can signify allegiance to a side of the debate between realism and anti-realism. An observable is a quantity produced by an experiment or an observation. There is no commitment to believe it corresponded to something that exists apart from the measurement or had a value before the measurement. Anti-realists use this term to emphasize that the quantities quantum physicists measure need have no existence apart from, or prior to, our observation of them. Realists use John Bell's term "beable" to refer to the reality that they believe exists whether we measure something or not.

Most scientific explanations, whether of the flights of cannonballs or of birds and bees, speak in terms of beables.

But not quantum mechanics! As Heisenberg and Bohr insisted, quantum mechanics speaks not in terms of what is, but only of what has been observed. There is, according to them, no useful talk about beables in the atomic domain; instead, quantum mechanics deals only in observables.

To measure an atom's observables, we impose on it a large, macroscopic instrument. By definition, that device is not part of the system whose observables we are studying. Nor is the observer.

Therefore, to be described in the language of quantum mechanics, a system must be part of a larger system that includes the observer and her measuring instruments. Hence our subsystem principle.

Most applications of quantum theory are to atoms and molecules or other tiny systems; in these cases the restriction is irrelevant. But some of us have the ambition to describe the whole universe. We feel that is the ultimate goal of science. However, the universe as a whole is not, by definition, part of a larger system. The subsystem principle frustrates our hope to have a theory of the whole universe.

There is a subtle but key difference between the idea that quantum mechanics is the theory of everything, and the hope of extending quantum theory to include the whole universe. What Professor Bernstein meant by his claim is that physics is the root of the correct description of everything—each considered as a subsystem of the whole. It is very different to imagine applying quantum theory to the entire universe, which would mean including us observers inside the system being studied, and our measuring instruments.

Over the last century several attempts were made to extend quantum mechanics to a theory of the whole universe. We will meet one of these later on; a part of our overall argument is that these attempts fail.

For one thing, making the observer a part of the system being described raises tricky questions of self-reference. It is not even clear that an observer can give a complete self-description, because the act of observing or describing yourself changes you.

But there are deeper reasons why quantum mechanics cannot be extended to a theory of the whole universe.

In several of my books (namely *The Life of the Cosmos, Time Reborn*, and *The Singular Universe and the Reality of Time*, written with Roberto Mangabeira Unger), I investigate the question of how physics may be extended to give a theory of the whole universe. I conclude that a theory of the whole universe must differ in several crucial aspects from any of the physical theories so far developed, including quantum mechanics. All these theories only make sense when interpreted as descriptions of a portion of the universe.

Indeed, the fact that quantum mechanics only makes sense when read as a theory of a part of the universe is, by itself, a sufficient reason for regarding quantum mechanics as incomplete. One thing we may ask of a theory that completes quantum theory is that it makes sense when extended to a description of the universe as a whole.

However, this is not the only line of thought that leads to the conclusion that quantum mechanics is incomplete. Other concerns and difficulties had far more influence on how the subject has evolved historically. For the time being, I will ignore the cosmological issues and focus on more immediate challenges.

THE PROCESS OF APPLYING general laws to a specific physical system has three steps.

First, we specify the physical system we want to study.

The second step is to describe that system at a moment of time in terms of a list of properties. If the system is made of particles, the properties will include the positions and momenta of those particles. If it is made of waves, then we give their wavelengths and frequencies. And so on. These listed properties make up the state of the system.

The third step is to postulate a law to describe how the system changes in time.

Before quantum physics, physicists had a simple but powerful ambition for science. At the second step we would be able to describe a system in terms that were complete, in two senses. Complete means, first of all, that a more detailed description is neither needed nor possible. Any other property the system might have would be a consequence of those already included. Additionally, the list of properties should be exactly what is needed to give precise predictions of the future. This is done using the laws. The future can be determined precisely, given our complete knowledge of the present. This is the second meaning of the description being complete.

Between Newton, in the late seventeenth century, and the invention of quantum mechanics in the 1920s, it was believed that the properties making up that complete description were the positions of all the particles and their momenta.

It might, of course, happen that we don't know the precise positions and momenta of all the particles making up a system. The air in this room consists of around 10^{28} atoms and molecules, so a complete listing of their positions is impossible. We have to use a very approximate description in terms of density, pressure, and temperature. These refer to averages of the atoms' positions and motions. Our bulk description will have to employ probabilities, and the predictions it makes will then be to some degree uncertain.

But the use of probabilities is just for our convenience, and the resulting uncertainties just express our ignorance. Behind our bulk description of a gas in terms of density and temperature, we continue to believe there is a precise description, which includes listing where every last atom is and how it is moving. We share a faith that if we had access to that description we could use the laws to predict the future precisely. That faith is based on the belief in realism—that there is an objective reality, which it is possible for us to know.

Quantum mechanics blocks this complacent ambition, because its first principle asserts we can know, at most, only half the information that would be needed to realize it.

THE COMPLETE INFORMATION NEEDED to precisely predict the future is called a *classical state*. "Classical" is how we refer to physics as it was between Newton and the discovery of the quantum. It is then natural to call a specification of half of that information a *quantum state*. The half is arbitrary; it can be chosen to be only the momentum, or only the position, or some mixture of these, as long as half the information needed to precisely predict the future is present, and half is missing.

The quantum state is a central notion in quantum theory. A realist will want to ask: Is it real? Does a particle's quantum state correspond precisely to the physical reality of that particle? Or is it just a convenient tool to make predictions? Perhaps the quantum state is a description, not of the particle, but only of the information we have about the particle?

We are not going to resolve these questions here. Experts disagree about them. We will soon enough have the chance to focus on these and other questions about the meaning and correctness of quantum mechanics. For now we take a pragmatic viewpoint and regard the quantum state as a tool for making predictions about the future.

A quantum state is a useful tool because it can do just that. This is our next principle:

> ***Given the quantum state of an isolated system at one time, there is a law that will predict the precise quantum state of that system at any other time.***

This law is called *Rule 1*. It is also sometimes called the Schrödinger equation. The principle that there is such a law is called *unitarity*.

Thus, while the relation between the quantum state and the behavior of an individual particle can be statistical, the theory is deterministic when it comes to how the quantum state changes in time.

As we said, quantum states with definite values of energy and momentum are represented by pure waves with exact frequency and wavelength. But these quantum states are very special. What about other quantum states, whose momenta are uncertain, so that they do not vibrate at a single frequency and with a single wavelength? More general quantum states are represented by waves with arbitrary profiles. These are sharp in neither position nor momentum, so if either of these quantities is measured, there will be uncertainties.*

There are also states of definite position and completely indefinite momentum; if we graph them, they look like spikes, which are zero everywhere except the single point where the particle is. Other states are peaked over a region of space and correspond to particles which are localized imprecisely, so we know only approximately where they are.

One way to make a general quantum state is by adding together pure waves, each with a different frequency and wavelength.

* When a wave represents a quantum state, we sometimes call it a wave function.

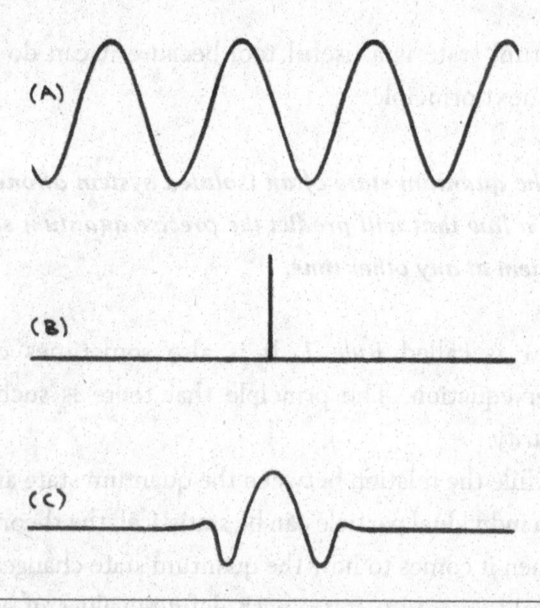

FIGURE 2. Three wave functions are illustrated showing how different kinds of states are represented. (A) shows a pure wave of a single wavelength, which corresponds to a definite momentum. The position is completely uncertain, as is required by the uncertainty principle. The spike in (B) shows a state with a definite position, but the wavelength is completely indefinite and uncertain. The intermediate case (C) is built by combining several wavelengths, so the momentum and position are both somewhat uncertain.

If we measure the energy of such a combination, we get a range of values corresponding to the different frequencies that make up the wave.

If this were music, the waves would be sound waves. A pure wave with a single frequency sounds a single note. Playing several notes simultaneously produces a chord. There is no limit to how many notes you can play at once, nor to how many quantum states can be added together.

Combining two states by adding the waves that represent them is called *superposing the states*. It corresponds to combining two ways the particle may have traveled to arrive at the detector. Earlier, when we divided cat and dog people into the living room and the kitchen, each room represented a quantum state, defined by a definite pet preference. When we brought everyone together in the dining room we superposed those two states.

This is an example of a general principle called the *superposition principle.*

> ***Any two quantum states may be superposed together to define a third quantum state. This is done by adding together the waves that correspond to the two states. This corresponds to a physical process that forgets the property that distinguished the two.***

Logically, a superposition of two states C and D communicates C *or* D. The person could be a dog lover *or* she could be a cat lover. The connector "or" means something has been forgotten. Someone might have been a C or they might have been a D, but when we've forgotten which, we can only say they are a C *or* a D.

As I have emphasized, quantum states are important because they evolve in time according to a definite rule. The relation between the quantum state and an observation is probabilistic, but the relationship between that state now and the quantum state at a different time is definite. But there is an important caveat, which is that the definite evolution rule applies only to systems that are isolated from the rest of the universe. Only in cases where the system is free from disturbances or influences from outside sources is the evolution rule deterministic.

When we make a measurement on a system, we disturb it, typically by forcing it to interact with a measuring instrument. So Rule 1 does not apply to measurements. This is true not only of measurements, but of any interaction between the system and outside forces. So is there anything special about measurements?

Measurements are special because they are where probabilities enter quantum theory.

Quantum mechanics asserts that the relationship between the quantum state and the outcome of a measurement is probabilistic. Generally, there is a range of possible outcomes of a given measurement. These will each occur with some probability, and these probabilities depend on the quantum state. In the case where we measure the position of a particle, this dependence is particularly simple:

The probability of finding the particle at a particular location in space is proportional to the square of the height of the corresponding wave at that point.

This is called the Born rule, after Max Born, who proposed it.

Why the square? Probability is always positive, but waves generally oscillate between positive and negative values. But the square of a number is always positive, and it is the square that is related to probability. The important thing to remember is that the larger the magnitude, or height, of a wave, the more likely that you will find the corresponding particle there.

These last few points are key to how quantum mechanics works, so let me summarize them: The wave represents the quantum state. When we leave the system alone, it changes in time deterministically, according to Rule 1. But the quantum state is only indirectly related to what we observe when we make a measurement, and that relation is not deterministic. The relation between the quantum state and what we observe is probabilistic. Randomness enters in a fundamental way.

But, even if the quantum state gives us only probabilities for what we observe, once we get a result, there is something that is definite, because afterward you know exactly what the state is. It is the state corresponding to the result obtained by the measurement. Suppose we measure an electron's momentum, and get the result that the electron is moving north with momentum 17 (in some units). Then, just after the measurement we know that the quantum state is NORTHWARD, MOMENTUM = 17.

This is enshrined in a second rule,* which we call *Rule 2*:

> ***The outcome of a measurement can only be predicted probabilistically. But afterward, the measurement changes the quantum state of the system being measured, by putting it in the state corresponding to the result of the measurement. This is called collapse of the wave function.***

For example, in our story about political and pet preferences, as soon as a person answers a question about either one, they go into the quantum state defined by having that definite preference.

Since the outcome of the measurement is probabilistic, so is the change in the quantum state dictated by Rule 2.

Once the measurement is over, the system can be considered to be isolated again and Rule 1 takes over, until the next measurement.

Rule 2 raises a whole bunch of questions.

> Does the wave function collapse abruptly or does it take some time?
>
> Does the collapse take place as soon as the system interacts with the detector? Or only later, when a record is made? Or perhaps later still, when it is perceived by a conscious mind?

* But you knew as soon as I mentioned Rule 1 that there had to be a second one. I should point out that in some textbooks, Rules 1 and 2 are switched.

Is the collapse a physical change, which means that the quantum state is real? Or is it just a change in our knowledge of the system, which means the quantum state is only a representation of that knowledge?

How does a system know a particular interaction has taken place with a detector, so that it should then, and only then, obey Rule 2?

What happens if we combine the original system and the detector into a larger system? Does Rule 1 then apply to the whole system?

These questions are all different aspects of the measurement problem.

Diverse answers have been given, which have been a source of controversy for nearly a century. We will have a lot to say about all this, once we have the full picture.

FOUR

How Quanta Share

Useful as it is under everyday circumstances to say that the world exists "out there" independent of us, that view can no longer be upheld.

—JOHN ARCHIBALD WHEELER

The superposition of quantum systems poses a grave challenge to realism. But an even more insidious set of obstacles to realism comes from how quantum mechanics describes systems which are built by combining simpler systems.

Superposition is about combining different possible states of a single system. As I said, it corresponds to "or." Quantum mechanics also has interesting things to say about combining two different systems to make a composite system. Suppose we have an electron and a proton. Each has to begin with its own quantum state. We can combine them to make a hydrogen atom. The whole atom has its own quantum state, which is made by combining the states of its constituents. This corresponds to "and." Each quantum state represents half the possible information needed for a complete description of its components. The joint quantum state also represents half the possible information about the atom. This leads to very interesting new phenomena.

Let us consider again people with two incompatible properties, political views and pet preference. Let's suppose Anna and Beth share an apartment. They talk about getting a pet. Individually, Anna is a cat lover and so is Beth. The state of their couple is just the combination of these. Each has a definite pet preference, so each has indefinite political views. If asked for her political preference, each will have a 50 percent chance of answering left and a 50 percent chance of answering right. So, if asked about politics, half the time they will discover they agree and half the time they will discover they disagree. In the state in question, in which they each separately have a definite pet preference, their political views are random and uncorrelated. Anna stating her political views has no effect on Beth's views.

Quantum physics also allows us to define states for the couple in which all their individual views are indefinite, but we can have definite knowledge of how their views relate. An important example of such a state is one in which the only thing that is certain is that, if we ask Anna and Beth the same question, they will disagree. This state is called CONTRARY. In this state you can ask them both any question, and whatever one asserts, the other will assert the opposite. Yet it is impossible to predict their individual answers.

CONTRARY is an example of a surprising phenomenon, which is that quantum states exist for two particles in which we know something about how the particles are related to each other, but nothing about each particle individually. We call such states *entangled*. The phenomenon of entanglement is something new, which comes into physics with the quantum and has no classical analogue.

The information that they will disagree, whatever question

they are asked, adds up to exactly half the information that would be needed to predict their actual answers. The other half is about their individual responses. So in the CONTRARY state, we know nothing about their individual views, and everything about how their views correlate. Hence, when in the CONTRARY state, Anna and Beth share a property which is not just the sum of properties they have individually.

The couple spend the evening together and wake up in the CONTRARY state. They each go off to work. Over lunch Anna's colleagues will ask her about either politics or pets. They decide only at the last minute which question to ask. Afterward they record which question was asked of Anna and what she said. Beth's colleagues do the same. This is repeated every day for a year, after which the two sets of colleagues meet at a conference and compare notes. What do they discover?

Half the time, Anna and Beth will have been asked different questions. Let's ignore these cases and look only at the days when they were asked the same question. In 100 percent of these cases, their answers disagreed with each other. This is in spite of the fact that, looked at individually, each of their answers appears to have been completely random.

As I've described this, it would not be hard to explain. All that is needed is that each morning over breakfast the couple toss a coin to decide who will give which answers, if asked. But there are analogous stories in which we study pairs of photons, rather than pairs of people. We can put pairs of photons in the CONTRARY state and measure various properties of them. Whenever we ask each the same question, they disagree. But we can show that this cannot be explained by any agreement established in advance of our asking. This was proved in an important paper, written by the Irish physicist John Bell, in 1964.

In the case of photons, we ask not about political or pet preferences, but about polarization. An electromagnetic wave consists of oscillating electric and magnetic fields. The oscillations are perpendicular to the direction in which the wave is traveling. These oscillations define a plane, which jumps around as the fields oscillate. We say that light is polarized when the electric field oscillates steadily in a particular plane. Individual photons that pass through a polarized lens, such as are common in sunglasses, have a well-defined polarization.

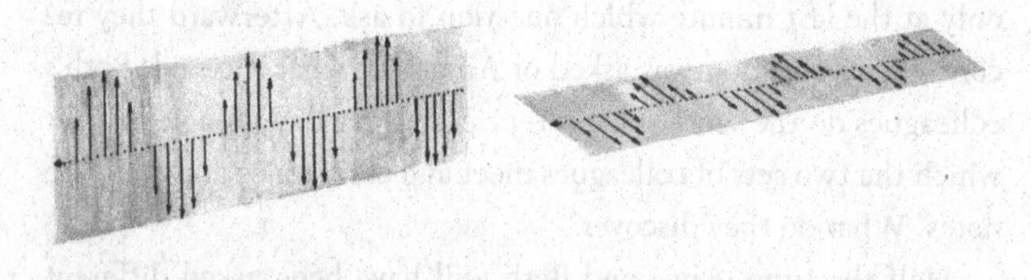

FIGURE 3. This figure shows what we mean by saying that electromagnetic radiation can be polarized. Here are traces of two waves moving through the electric field, in the absence of external currents and charges. Note that the electric field points perpendicular to the direction of motion of the wave. The oscillations of the field, together with the direction of motion, define a plane in three-dimensional space. This is called the plane of polarization. We show two planes of polarization, perpendicular to one another.

We can produce pairs of photons that together have polarizations in the CONTRARY state. To show this we let them travel in opposite directions till they are far from each other, then we put in their paths polarized glass, which they either pass through or not. In the state CONTRARY, if the glasses have the same plane of polarization, one of the two photons will pass through the glass, but the other won't. Which one passes, however, is random because in the state CONTRARY their individual properties are completely uncertain.

We can also swivel one glass, which rotates the plane of polarization to one side. The two polarizers are now at different angles. Now some of the time both photons pass. How frequently both pass depends on the angle between the polarizers. When the angle between the two polarizers is zero, we are asking the same question on each side and it never happens that both pass. Let us then rotate one polarizer a bit, so that they are asking slightly different questions. Now in a few instances photons pass on both sides. We ask about how the proportion of cases in which they both pass increases as we vary the angle between the two planes of polarization.

Bell introduced an assumption which expressed the idea that physics is local, in that information cannot travel faster than light. This requires that when the two photons are very far apart, the questions I choose to ask one photon cannot affect the answers the other will give.

From this assumption, Bell derived a restriction on the proportion of cases in which both photons pass their polarizers. This restriction depends on the angle between the two planes of polarization.

Bell asked first whether the restriction is violated by the predictions of quantum mechanics. He found that for certain angles it is violated. This means that quantum mechanics violates Bell's principle of locality. We can easily see that this is the case in the story of our couple. When Anna and Beth each go off to work, they share a single quantum state, the state CONTRARY. This is not a property of either of theirs as individuals. It is a shared property; it makes sense only when it is ascribed to the couple. This situation is already in tension with the philosophy that physical properties are local.

But it gets worse. When Beth's coworkers ask her about her pet preference she says she loves cats. This immediately changes her quantum state, as prescribed by Rule 2. It was originally indefinite, but now she is purely a cat person. If asked again about pet preference she is certain to say "cat," so the state CAT defines her.

But by the same logic, because they started the day in the CONTRARY state, Anna became at that moment a person with a definite preference for dogs. If asked by her colleagues which pets she prefers, Anna is now 100 percent certain to say "dogs."

Thus the measurement of Beth's preference appears to instantly affect Anna's state. In spite of the fact that it was Beth who was measured, and Anna has talked to no one, Rule 2 applies to Anna as well. This is an example of the phenomenon known as quantum nonlocality.

The story would be exactly the same if Beth were asked her political leanings. Whichever way she answered, Anna would instantly become the other.

Once Beth is asked about one of her preferences, she and Anna no longer share a state. Beth now has a definite state of her own, and you can say this was the result of her being measured. What is weird is that, because they were originally together in the entangled state CONTRARY, when Beth is queried this immediately changes Anna's state as well. By virtue of the answer that Beth gives, Anna is immediately defined as being in a quantum state of her own, to wit, the opposite of whatever answer Beth gives.

This happens even though no one has yet asked Anna anything. Beth and her colleagues may be light-years away, so no information about what Beth was asked and what she answered could reach Anna for years, assuming the usual restriction on the transmission of information. This is to say that Anna herself cannot know yet that her quantum state has changed. But it has, if quantum theory is correct.

Of course, the story would be the same if it were Anna who had been asked first. The consequences of sharing an entangled state are entirely symmetric.

THE STRANGE BEHAVIOR of the quantum state CONTRARY was discovered by Einstein, and it was the centerpiece of a paper he wrote in 1935 with two younger colleagues, Boris Podolsky and Nathan Rosen.[1] The three authors (sometimes abbreviated as EPR) used an experiment like I've described to argue that quantum mechanics must be incomplete. To arrive at that conclusion they gave a criterion for when a property of a physical system must be considered real. Here is their criterion:

> If, without in any way disturbing a system, you can determine a property of it with 100 percent certainty, there must be an element of physical reality associated to that property.

Einstein and his collaborators also assumed that you could only disturb a system by doing something physically to it. Most importantly, they also assumed that any physical disturbance is local, and is hence restricted to traveling at the speed of light or less. This implies in particular that

> Anna cannot be physically affected by the choice of questions Beth is asked until enough time has passed for a light signal to have carried the information about which question Beth was asked from Beth to Anna.

We have just seen that once Beth's colleagues query Beth about her pet preference, they also know Anna's pet preference.

However, Einstein and his friends believed strongly in the principle of locality, which implies that, because they are far apart, Anna cannot have been disturbed by questions asked of her faraway friend Beth. Hence, the criterion for reality just enunciated is satisfied and we can conclude that *Anna's pet preference is an element of reality.*

Moreover, what is real concerning Anna can't be affected by anything that happens or doesn't happen to Beth. So Anna's pet preference must be real whether or not Beth's pet preference was queried.

Now, notice that Beth's colleagues might instead have asked about her politics. The same argument works and we must conclude that *Anna's political preference is also an element of reality.* And again, *this is true whether or not Beth's politics was queried.*

So we must conclude that *both Anna's pet preference and her politics are elements of reality!*

But quantum states cannot simultaneously describe both someone's politics and their pet preference. *Hence, Anna's quantum state incompletely describes her.*

And so, concluded Einstein, Podolsky, and Rosen, the description of the world in terms of quantum states is incomplete.

I've been thinking about this argument since my first year of college. So far as I can tell it's logically correct. But notice that it depends on the assumption that physics is local. Einstein and his young friends assumed locality when they posited that *Anna cannot be physically affected by the choice of questions Beth is asked when they are far apart.*

Bell made exactly the same assumption, regarding photons rather than people, in deriving his restriction.

When the two photons are very far apart, the questions I choose to ask one photon cannot affect the answers the other will give.

This is, indeed, the only non-trivial assumption in Bell's argument. So since, as I said, Bell's restriction disagrees with quantum mechanics, it must be that quantum mechanics itself disagrees with locality.

But we can go further and test directly whether locality, as assumed by EPR and Bell, is violated by nature.

The importance of Bell's restriction is that it applies not only to quantum mechanics. The restriction he derived constrains any theory that satisfies Bell's and EPR's principle of locality. This includes theories that are intended to replace quantum mechanics. It will apply equally to any theory which might be invented in the future. This means that we can set up experiments that directly test the locality principle.

Fortunately, Bell's restriction could be tested by a relatively inexpensive device, hand-built in a single room. A few brave souls began the work of building experiments to test the theorem. After several attempts got partial and contradictory results, the definitive experiments were carried out in Orsay, near Paris, in the early 1980s by Alain Aspect and his collaborators, Jean Dalibard, Philippe Grangier, and Gérard Roger.[2]

In Aspect's experiments the entangled particles are photons and the questions asked are about their planes of polarization. These experiments begin with an atom raised from its ground state into an excited state, by a photon from a laser. These are chosen so that when the excited atom decays back to the ground state, it does so in a way that produces a pair of entangled photons, in the state CONTRARY. The photons fly off in opposite directions and after a few feet encounter polarizers, which measure their polarizations relative to a plane. The plane of each polarizer can be set freely, in whatever position the experimenter chooses, so the correlations of the polarizations of the two photons can be measured. The results cleanly violated Bell's restriction while agreeing precisely with the predictions of quantum theory.

The experiments tell us that the assumption of Bell locality highlighted above is false! The quantum world does not obey the principle of locality.

If this is not the most shocking news you have heard from the world of science, you have perhaps not understood it. *Nature does not satisfy the idea of locality.* Two particles, indeed two objects in the world, situated far from each other, can share properties that cannot be attributed to properties separately enjoyed by either.

At this point it is natural to wonder if the principle that information cannot be transmitted faster than light could be violated, by taking advantage of the circumstance that Beth and Anna share an entangled state. Could the fact that Anna's state is changed abruptly, based on which question Beth is asked, be used by Beth's colleagues to send a message instantaneously to Anna's colleagues?

The answer is that information cannot be sent faster than light, because the relation between Anna's state and the answers she gives is random. No matter what question Anna is asked, her answers are 50 percent either way. This is true before Beth is queried, when she shares the state CONTRARY with Anna, and it remains true afterward. It is only when the lists of answers each gave to a series of questions are brought together and compared that evidence of mysterious correlations appears. And the lists are ordinary classical objects that cannot be transmitted faster than light.

There is another, related possibility, which Aspect and his colleagues could also test. Perhaps, at some deeper level than that described by quantum theory, the two atoms are in communication, so that the first photon to be measured transmits information to the other photon about what question it was asked. Then the principle of locality could still be satisfied. But now we have to reckon with special relativity, which maintains that no information can travel faster than light. To test for this possibility, the experiment was redone with a random switch on one side, which could very rapidly choose which question would be asked of its photon. This switch was fast enough that the choice was made while the photons were in flight. Thus the switching happened faster than could be communicated to the other photon by any signal traveling at light speed or less. The result was unchanged. If the two photons are in communication, their messages are being transmitted much faster than light, and relativity theory is violated.

What are we then to make of the argument of Einstein, Podolsky, and Rosen? As clever as it was, the argument must be considered, in light of the experimental findings, to be wrong, because it relies on an incorrect assumption, which is the assumption of locality. The experimental tests of Bell's inequality show us that, once Anna and Beth are entangled in the state CONTRARY, Anna in fact *is* physically affected by the choice of questions Beth is asked. This remains true even when they are far apart. This is true in quantum mechanics, and, the experiments imply, it must be true in any deeper theory that completes quantum mechanics.

Nevertheless, EPR's paper was enormously important, because it exposed an unexpected aspect of quantum physics, which was entanglement. This took decades to appreciate; indeed, the EPR paper was way ahead of its time. Apart from the discovery of entanglement, the paper was the starting point for Bell and hence for the shocking experimental discovery that physics violates the principle of locality.

Bohr, the great anti-realist, replied right away to the EPR paper, with an especially obscure example of his style of reasoning.[3]

He took issue with EPR's criteria for reality by pointing out that a measurement of one of the particles disturbs the other particle indirectly, by disturbing the context within which the properties of the other particle make sense.

For the next fifteen years there is just one paper written which cites the EPR paper. The next several citations are by Bohm and Everett in the 1950s. John Bell was just the sixth author to cite EPR, which he did in his great paper of 1964, almost thirty years later. Yet the paper was cited more than sixty times in 2015, and again in 2016. We now, finally, live in the era of entanglement.

In recent years, the sharing of properties among entangled pairs has been confirmed in experiments in which the pairs are separated by hundreds of kilometers. Entanglement is fast evolving from a laboratory curiosity into a technology. It is now considered a resource, which is at the heart of a new kind of computer—a quantum computer. In the near future entanglement may allow us to break codes long thought secure as it also makes possible new kinds of codes that are truly unbreakable. There are already in orbit quantum communications satellites, which employ entangled pairs to encrypt messages they transmit.

Einstein's first revolutionary papers appeared in 1905, when he was twenty-six. Thirty years later, the EPR paper was the last paper by Einstein to shake physics to the core. It is given to very few to lead science over three decades. Einstein never ceased trying to find the deeper theory beyond quantum mechanics, and two decades further on, he was still working in his notebook in the hospital the night he died. But he failed, and the simple reason was that he never understood that the central assumption behind many of his great papers—the principle that physics is local—was wrong.

There is no reason Bell's 1964 paper could not have been

written in the late 1930s, shortly after EPR. And the experimental disproof of locality could have happened shortly after. One can only wonder what Einstein would have thought had he learned of Bell and Aspect in the 1940s.

TOGETHER THE STORIES I have told so far illustrate the strangeness of the quantum world. They have taught us about the wave-particle duality, superposition, and the uncertainty principle.

Stranger still was how quantum properties can be entangled and shared among systems that are widely separated in space. This was the ultimate lesson of the story told by Einstein, Podolsky, and Rosen. But it was only in John Bell's retelling that the true moral of the story was revealed to be the radical nature of quantum nonlocality.

As we saw, superposition can be understood as a quantum version of "or," which I will indicate as ***or***. When we combine two systems, we use a quantum version of "and." I will write this as ***and***. Each behaves differently from the ordinary usage of "or" and "and" that we are used to from everyday life. But it is when they act together that truly strange things happen. We see this in a famous experiment called *Schrödinger's cat*.

Let us start with a very simple model of an atom, which can exist in two states: an excited, unstable state, which we call EXCITED, and a stable ground state with lower energy, called GROUND. EXCITED, being unstable, will decay into GROUND by emitting a photon, which carries away the energy. These decays take place at a rate measured by the half-life of the excited state.

Let us put an atom in the state EXCITED in a box and wait a time comparable to the half-life. If we don't look in the box, we can deduce only that the probability is about half that if we open the box we will see that the atom has decayed to the state GROUND. But what is the state before we look inside the box? According to quantum mechanics, it is neither EXCITED nor GROUND, but a superposition of them. We can write this as

ATOM = EXCITED or GROUND

According to Rule 2, this superposition has the potential of becoming, when we look, either of the two states: EXCITED or GROUND. If we have a large collection of such states, then we can determine probabilities for each of these outcomes. These probabilities change in time. Just after making the atom, the probability that it has decayed is very small. Many times the half-life later, it has almost certainly decayed.

A superposition is not the same as having one or the other state with varying probabilities. One reason is that when we make the energy uncertain by superposing two states of different energies, another observable will be made certain. This is like the way we made our visitors have definite political views by superposing their states with different pet preferences. So we can always find a question complementary to the energy that the answer to will be YES with certainty. That would not be the case if we were just dealing with the probabilities of being EXCITED or GROUND.

We next put a Geiger counter in the box, and set it up to send out a pulse of electricity whenever it sees a photon.

From the point of view of quantum mechanics, the Geiger counter can also exist in two different states. There is the state NO, in which it hasn't seen a photon, and the state YES, when it has. It can also exist in superposition of these two states.

We put the atom in the box with the Geiger counter. We must

be careful to set them up so that initially the atom is in the state EXCITED and the Geiger counter is in the state NO.

INITIAL = EXCITED **and** NO

By ***and*** we understand that these states, being states of two different systems, are being combined, not superposed.

Much later, if everything is working well, we expect to see the atom in the state GROUND and the Geiger counter in the state YES. This corresponds to the Geiger counter having detected the photon emitted when the atom decayed.

FINAL = GROUND **and** YES

In between, the system is in a superposition of these two states.

IN BETWEEN = (GROUND **and** YES) **or** (EXCITED **and** NO)

The total system is a superposition of a state where the atom is in the undecayed state EXCITED *and* the Geiger counter is in the state NO with the other possibility, which is the state in which the atom has decayed to GROUND *and* the Geiger counter is in the state YES, in which it has seen the photon.

This state IN BETWEEN is an example of a *correlated state.* We call it that because the properties of the two systems are correlated. The state of the atom is uncertain, but if we know what state the atom is in, we can deduce which state the Geiger counter will be in.

But if we then open the box and look inside, we never see a superposition. Looking inside is a measurement which is governed by Rule 2. We see either that the Geiger counter has clicked, so the atom has decayed, or that the atom is still excited and the counter has yet to click.

This seems downright weird. Here are some of the questions it raises.

Why are there two rules for how quantum systems change in time, rather than one?

Why do we treat measurements and observations differently from other processes? Certainly a measurement device is just a machine made out of atoms. Shouldn't there just be one rule for how things change in time, which applies in all cases?

And just what is it about measuring devices that makes them different? Is it just the size or complexity of the device? Is it the vast number of atoms making it up? Or is it the fact that it can be used to gain information?

When does the collapse to a definite state happen? Is it when the atom meets the detector? Or when the signal is amplified? Or is it not until we become conscious of the information?

These questions are all aspects of the measurement problem.

The simplest answer is that, one way or another, it must be this way. We never observe large things to be indefinite: in our world there are no Geiger counters that both have and have not clicked. Every question we ask has a definite answer. But we need superpositions to explain atoms and radiation.

To emphasize how strange this all is, Schrödinger put a cat in the box along with the atom and the Geiger counter. He wired up the signal from the Geiger counter to a transformer, whose output was clipped to the cat's ears. When the Geiger counter signaled its detection of the photon, the cat got a fatal pulse of electricity.

(Of course Schrödinger didn't actually do this. This is a thought experiment intended to shock us, not the cat.)

We wait a half-life and then open the box. Do we apply Rule 1 or Rule 2? Let's discuss what each of the two rules would predict.

Assume first that Rule 1 applies to the whole system inside the box, including the cat. That system consists of the atom, the Geiger counter, and the cat. There are again two states with easy interpretations. One of these is the initial state

INITIAL = EXCITED **and** NO **and** ALIVE

This is the state in which the atom is excited, the Geiger counter has detected nothing, and the cat is alive. After a long time we can be sure the atom has decayed and the cat has died.

FINAL = GROUND **and** YES **and** DEAD

This state is the result of the decay and features a stable atom in the ground state, a detector that clicked, and a dead cat.

In between, the state is a superposition of these two possibilities.

IN BETWEEN = (EXCITED **and** NO **and** ALIVE) **or** (GROUND **and** YES **and** DEAD)

But a cat is a mammal, with a brain and perhaps a conscious mind. It is nearly as complex as we are. So why does it make sense for the cat to be in a superposition of alive and dead? If it doesn't make sense for us to exist in a superposition, it surely doesn't for the cat either. If we apply Rule 2 to our observation, we should also apply it to the cat, who in essence observes the signal from the detector.

So we'd better apply Rule 2. When we open the box, the system makes a choice and jumps into a definite state. We find either a live cat or a dead cat.

So Rule 1 alone does not apply to humans or cats. But does it apply to Geiger counters? And where is the line? Why does it apply to atoms and not to big collections of atoms like detectors, cats, and humans?

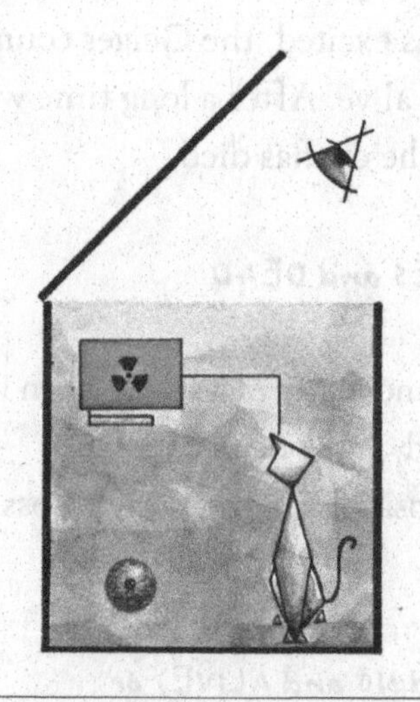

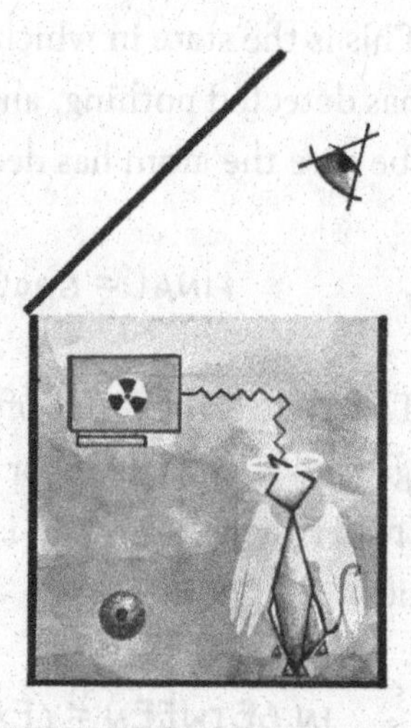

FIGURE 4. The Schrödinger's cat thought experiment. A detector is constructed to respond with an electrical pulse to a photon that would be emitted by an atom decaying and jumping down from an excited state into the ground state. The cat is connected to the circuit, so when the pulse comes it will electrocute him. After a short time, the atom is in a superposition of its excited and decayed states. Rule 1 applied to this case predicts that the cat inside the closed box is then in a superposition of two states, alive and dead.

This conversation is called the puzzle of Schrödinger's cat. A measure of the fecundity of the human imagination is the number of responses that have been offered to this puzzle.

A FEW YEARS AFTER Bell published his restriction, an even more powerful result was published which further limits the options for realist quantum theories. To describe it we go back to Bell.

One way to put what is surprising about Bell's result is that the answers that Anna gives to the questions she is posed have to depend on the questions that Beth is asked. This is shocking because Anna and Beth are separated, so locality would preclude such a dependence. But notice that they don't need to be separated for the conclusion to apply. Then the result that Anna's answers depend on the questions Beth is asked is surprising for another reason.

Earlier we talked about pairs of measurements which are mutually incompatible, like a particle's position and momentum. In these cases it seems that the act of measuring one quantity interferes with or disturbs the value of the other. We described this by saying that the order in which the two measurements are made matters.

But notice that the case of Anna and Beth is not like this. Questions asked of Beth are completely compatible with questions asked of Anna. The order in which they are questioned doesn't matter. This was true when the two friends were far apart when they were questioned, but it would remain true if they were standing next to each other.

Still, even if the order in which we question the two friends is irrelevant, so that questions to one are compatible with questions to the other, it remains the case that the answers Anna gives depend on the choice of which questions Beth is asked.

This dependence is called *contextuality*, because the answers Anna gives turn out to depend on the overall context, even to the point that they depend on choices made about which other questions will be asked. It turns out to be widely true of quantum mechanical systems. Contextuality occurs in situations in which our system is described by at least three properties, which we can call A, B, and C. A is compatible with both B and C, so A may be measured simultaneously with either B or C. But B and C are not compatible with each other, so we can measure only one at a time.

So we can measure A and B or we can measure A and C. We make a series of experiments in which we make both choices, and we record all the answers. When we do we will find—assuming that quantum mechanics is correct—that the answers to A depend on whether we chose to measure B or C along with A. The conclusion is that nature is contextual. This is the case with quantum mechanics, and experiments have been done which confirm this prediction of the theory. So it must be true in any deeper theory which will replace quantum mechanics.

This result was first proved by John Bell in the early 1960s, before he published his result on nonlocality. He submitted it to a journal but the paper was apparently lost for two years, "on the editor's desk," so it wasn't published till 1966. By then the result had been proved again by two mathematicians, Simon Kochen and Ernst Specker, so the result that quantum mechanics is contextual is often attributed to them, but it ought properly to be called the Bell-Kochen-Specker theorem.[4]

Quantum mechanics was invented in order to explain certain puzzling experimental results concerning light, radiation, and atoms. The three new phenomena we discussed in this chapter—entanglement, nonlocality, and contextuality—are a far distance more puzzling. Each is so weird that they were for a time used to argue that quantum mechanics must be wrong, till experiments confirmed that they are indeed all aspects of the natural world.

This was certainly not anticipated. Entanglement, nonlocality, and contextuality each emerged from the study of quantum systems, and it is very fair to say that they were each predictions of the quantum theory which, very surprisingly, turned out to be true.

These three aspects of quantum physics present severe challenges to realism. Indeed, they rule out large classes of realist theories. In particular, nonlocal entanglement is incompatible with all theories whose beables influence each other only through local forces, whose actions propagate at the speed of light or slower. Any realist theory which can mimic quantum mechanics must then describe a world which violates this condition and so openly embraces nonlocality. This is why Einstein talked of "spooky action at a distance." The choice we face is simple: we may give up realism and accept quantum mechanics as the final word, or we can move ahead and seek to understand how nature violates locality while still managing to make sense at all.

FIVE

What Quantum Mechanics Doesn't Explain

Quantum mechanics doesn't answer every question we can ask about the atomic world, but it gets a lot right. This is a good time to sum up what we've learned about what quantum mechanics does and does not explain.

Roughly speaking, quantum mechanics predicts and explains two kinds of properties: properties of individual systems, and averages taken over many individual systems. These are very different.

When we can attribute a definite value to a quantity—as we can when we make a measurement—this is a property of the individual system that has been measured. But often the uncertainty principle forbids us from discussing anything other than averages.

To what do these averages refer? Because of the uncertainty principle it can happen that two atoms, prepared identically in the same initial state, give different values when measured later. For example, atoms prepared in the same starting position will tend to spread out, and be found in different places later. When the final answers vary we can still measure their average value. Quantum mechanics tells us these averages are taken over many runs of an experiment. An experiment requires us to prepare many copies of a system, wait and then measure each copy, and then take the average of the results.

A collection of atoms which are similar in some way but different in others is called an *ensemble*. Quantum mechanics deals with ensembles. These may be defined by fixing one quantity, such as energy, to have some definite value, while other parameters vary over a range of values, as required by the uncertainty principle. When we speak of averages or probability in quantum mechanics, we are usually referring to something that can be measured by taking an average over the members of an ensemble consisting of many copies of the atom in question.

That is often easy to do because many experiments deal with a collection of atoms, such as a gas. These are real ensembles, because the atoms in the collection are real. Sometimes, though, the ensemble exists only in the theorist's imagination.

It is normal to explain the results of averaging over many copies of an individual system in terms of the properties of those individual systems. However, in quantum mechanics it is often the other way around, and a property of an individual atom will be explained in terms of averages over many atoms. But how can the collective determine the individual? These kinds of cases are at the heart of what is most mysterious about the quantum world.

One of the individual properties that quantum mechanics can discuss is the energy of an atom or molecule. It turns out that in quantum mechanics the energies of many systems come in certain discrete values, called the spectrum. The spectrum is a property of individual atoms, as it can be observed in experiments involving just one atom. Atoms, molecules, and various materials all have spectra, and in all these cases they are correctly predicted by quantum mechanics. More than that, quantum mechanics *explains* why these systems can have only these energies. It accomplishes this by making use of the wave-particle duality. This is one place where averages over many systems are used to explain what happens in an individual system.

The explanation involves two steps. The first is to use the relation between energy and frequency, which is the foundation of the wave-particle duality. A spectrum of discrete values of energy corresponds to a spectrum of discrete frequencies. The second step exploits the picture of a quantum state as a wave. A wave ringing at a definite frequency is like a bell or a guitar string producing sound. The string resonates when plucked, as does the bell when struck, ringing at a definite frequency.

We then use the equation for quantum states changing in time to predict the resonant frequencies of the system. The equation takes as input the masses of the particles involved in the system and the forces between them, and gives as output the spectrum of resonant frequencies. These are then translated into resonant energies.

This works well. For example, if we input that the system is made of an electron and a proton, bound together by their electrical attraction, the equation outputs the spectrum of the hydrogen atom.

In most cases, there is a state of lowest energy, which is called the ground state. States of higher energy are called excited states. You excite the ground state by adding the energy needed to bring it up to the level of one of these excited states. This causes the state to transition from the ground state to the excited state. The added energy is often delivered by photons. Excited states tend to be unstable, because they can drop back down to the ground state by radiating away the excess energy in the form of a photon. The ground state has no state below it to decay to, and so it is stable. Most systems spend most of the time in their ground states.

This method has been tested on a great many systems, including atoms, molecules, nuclei, and solids. In all cases the predicted spectra are observed. In addition to getting the spectrum of possible energies right, quantum mechanics makes predictions for averaged quantities, such as average values of the positions of the particles making up the system.

For each resonant frequency, the equation that defines quantum mechanics can be solved to yield the corresponding wave. We then use Born's rule (that the square of the wave is proportional to the probability of finding the particle) to predict probabilities for the particle to be found different places.

The states of definite energy have indefinite positions. Suppose we prepare a million different hydrogen atoms, all in the ground state. In each of these, we measure the position of the electron (relative to the proton, which is held fixed in the center of the atom). Each individual measurement results in a different position. Measuring a million different atoms gives us a million different positions. Some will be far from the proton, but most will be clustered around the proton in the center. The array of possible positions makes up a statistical distribution and it is this distribution, rather than a definite position, that quantum mechanics predicts.

According to the uncertainty principle, the position of any one of the electrons cannot be predicted. But the statistical distribution of positions, which results from measuring a great many cases, can be found. These statistical distributions are computed by squaring the wave.

To summarize, quantum mechanics makes two kinds of predictions. It makes predictions for the discrete spectra of energies, or other quantities, a system can have. And it also makes predictions for statistical distributions of quantities such as positions of particles.

In every case I know of, these two kinds of predictions have been confirmed by experiment. This is exceedingly impressive.

But does quantum mechanics explain how individual atoms work? Is a successful prediction always the same as an explanation?

IT IS EQUALLY IMPRESSIVE what quantum mechanics does not do. It does not describe or predict where a particular individual electron will be found. Because it deals in averages, quantum mechanics has little to tell us about what goes on in individual systems.

There are lots of cases where we deal with averages. We have no problem measuring the average height of Canadians. This is because each Canadian is some definite number of centimeters tall. We add all those centimeters up, divide by the number of Canadians we measured, and we get the average.

In cases like this, the average is made up of individual heights, which are properties of individuals. We could choose to work with the whole list of heights, but for many purposes, such as designing furniture or cars, the averaged value is all we need. If we need anything else, it is likely to be the *standard deviation*, which tells us the typical range of variations of height. Using the average and standard deviation, an airline could (if it wanted to) build airplane seats in which 95 percent of Canadians would be comfortable.

In these cases, the information which we ignore when we use averages is really present in the world, but we choose to suppress it in favor of the averages. The uncertainties which arise from our use of probabilities are purely due to our ignorance.

But suppose that each time we measured someone's height, we

got a different result. There is then an element of genuine randomness, because there is no way for us to know how tall someone might be the next time they are measured. That is closer to the case we deal with in quantum theory. What does the average signify, and what does it explain, when there is no story about individual cases?

Quantum mechanics makes correct predictions for averages, in spite of having nothing definite to say about individual cases. We seem to lack the kind of explanation we usually expect in cases like height, where the basis of an average is found in the fact that the average is composed of individual cases.

One of the most unexpected aspects of quantum mechanics is that a system can change over time in two ways. I described these in chapter 3. Most of the time the quantum state evolves deterministically under Rule 1. But when we make a measurement of the system it evolves in a very different way under Rule 2. The measurement will produce one number out of a range of possible values. Just after the measurement, the quantum state jumps into a state corresponding to the definite value which was measured in the experiment.

Rule 1 is continuous and deterministic; Rule 2 by contrast is abrupt and probabilistic. The state jumps abruptly just after the measurement, but quantum mechanics predicts only probabilities for the different outcomes, and hence for which state the system jumps to.

Most people are perplexed when they learn about these two rules. As we discussed before, the situation is genuinely puzzling. The first thing that puzzles them is the measurement problem:

What's so special about a measurement? Aren't measuring devices and the people who use them made of atoms, to which Rule 1 applies?

Rule 1, by dictating how a quantum system changes in time, plays the same essential role in the theory that Newton's laws of motion played in pre-quantum physics. Like Newton's laws, Rule 1 is deterministic. It takes an input state and evolves it to a definite output state at a later time. This means it takes input states which are constructed as superpositions to output states which are similarly constructed from superpositions. Probability plays no role.

But measurements, as described by Rule 2, do not evolve superpositions to other superpositions. When you measure some quantity, like pet preference or position, you get a definite value. And afterward the state is the one corresponding to that definite value. So even if the input state is a superposition of states with definite values of some observable quantity, the output state is not, as it corresponds to just one value.

Rule 2 does not tell you what the definite value is; it only predicts probabilities for the different possible outcomes to occur. But these probabilities are not spurious; they are part of what quantum mechanics predicts. Rule 2 is essential, because that is how probabilities enter quantum mechanics. And probabilities are essential in many cases; they are what experimentalists measure.

However, quantum mechanics requires that Rule 1 and Rule 2 never be applied to the same process, because the two rules contradict each other. This means we must always distinguish measurements from other processes in nature.

Yet if we are realists, then measurements are just physical processes, and there is nothing special that should distinguish them fundamentally from anything else that happens in nature. Thus, it is very hard to justify giving a special role to measurements within realism. Hence, it is hard to square quantum mechanics with realism.

AT THE END OF THE DAY, the question will be this: Can we live with these contradictions and puzzles, or do we want and expect more from science?

SIX

The Triumph of Anti-Realism

Quantum theory does not describe physical reality. What it does is provide an algorithm for computing *probabilities* for the macroscopic events ("detector clicks") that are the consequences of our experimental interventions. This strict definition of the scope of quantum theory is the only interpretation ever needed, whether by experimenters or theorists.

—CHRIS FUCHS AND ASHER PERES

The person who first understood that quantum physics would require a radically new theory based on a duality of waves and particles was Albert Einstein. Einstein was a realist to the core. Yet the quantum revolution he sparked culminated twenty years later in a theory that requires that measurements be singled out and treated differently than all other processes—a distinction that, as I discussed in the last chapter, is foreign to realism. The resolution, according to most of the pioneers of the quantum world, was to give up realism. How did this abandonment of realism come to happen?

The idea of a duality of wave and particle first appeared in Einstein's studies of the nature of light in the early years of the twentieth century. By that time physicists had considered theories in which light is a particle and theories in which light is a wave, but always one or the other. Newton considered the wave theory and rejected it in favor of a theory in which light is conveyed by a stream of particles traveling from objects to the eye. (Some ancient thinkers had them going the other way, which led to trouble explaining why we don't see in the dark.) Newton's reason for this choice was interesting: he thought that particles did a better job of explaining why light travels in straight lines. Waves, he knew, could bend as they diffract around obstacles, and he didn't think light could do that. Newton's particle theory of light reigned until an English scientist named Thomas Young showed in the early years of the nineteenth century that light did indeed bend and diffract at the edges of obstacles and as it passed through slits. Young was a medical doctor who contributed to several areas of science and medicine as well as Egyptology. He was an expert in a broad range of fields, something that the rapid expansion of the sciences was shortly to make impossible. He was sometimes called "the last person to know everything," but his greatest accomplishment was his wave theory of light, which, together with the experimental evidence he provided for diffraction, led to the overthrow of Newton's particle theory.

One of the examples Young considered was the double slit experiment, which is illustrated in figure 5. Water waves originating from the left pass a breakwall broken by two slits, on the way to a beach on the right. The waves from the two slits interfere with each other: the height of the water at each point to the right of the wall is a combination of waves propagating from the two slits. When the peaks of the two waves coincide, you see reinforcement—the combined wave is at its highest; but when the peak of one wave arrives in coincidence with the trough of the other, they cancel each other out. The result is the pattern graphed at the right, which is called an *interference pattern*. The key thing to understand and remember is that the interference pattern is the result of waves arriving from the two slits.

Thomas Young was able to construct the analogue of a double slit apparatus for light, and he saw an interference pattern. This made a strong case for light being a wave.

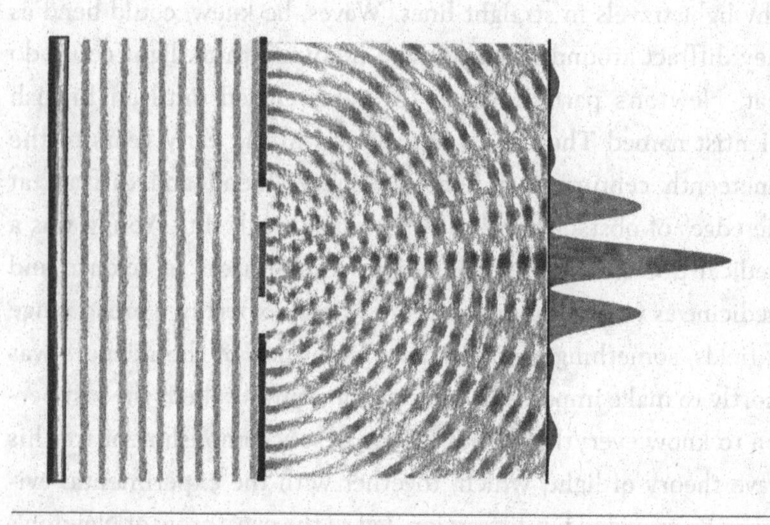

FIGURE 5. The double slit experiment, which shows that light behaves as a wave.

Further support for the idea that light is a wave came from the Scottish physicist James Clerk Maxwell, who showed around 1860 that light is a wave shimmying through the electric and magnetic fields that fill space as they convey forces between charges and magnets.

Einstein accepted Maxwell's hypothesis but added one of his own, which was that the energy carried by light waves comes in discrete packets, which he called photons. Thus was born the idea that light has a dual nature—it travels like a wave but conveys energy in discrete units like a particle. Einstein tied together the waves and particles by a simple hypothesis, according to which the energy a photon carries is proportional to the frequency of the light wave.

Visible light spans a range of frequencies, within which red light has the lowest frequency. Blue light is almost the highest frequency we can see, vibrating roughly twice as fast as red. Thus, a blue photon carries roughly twice the energy of a red photon.

What led Einstein to make such a radical proposal? He knew of experiments which could distinguish the effect of increasing the intensity of a beam of light from the effects of changing its color or frequency. This was done by shining light on metal, which caused some of the electrons in the metal to jump out, making an electric current that could be detected by a simple instrument an electrician might use.

The experiments measured how much energy the jumping electrons acquired from the light shining on the metal. The results showed that if you want to increase the energy each electron gets, you have to turn up the light's frequency. Dialing up the intensity has little or no effect; this merely raises the number of photons falling on the metal, without changing the energy the electron acquires from individual photons. This accords with Einstein's hypothesis that the electrons take energy from light by absorbing photons, whose energy is each proportional to the light's frequency.

Electrons are normally imprisoned in a metal. The energy a photon gives to an electron is like atomic bail: it liberates the electron, allowing it to travel free of the metal. But that bail is set at a certain amount. Photons which carry too little energy have no effect. If the electron is to escape, it has to get its energy from a single photon; it cannot collect up a lot of small increments. Hence, red light doesn't suffice to get a current started, but even a few photons of blue light will liberate some electrons, because each photon carries enough to bail out an electron.

The fact that no amount of red light, no matter how intense, will suffice to liberate an electron, while even a tiny amount of blue light succeeds, was to Einstein a big hint that the energy of light is carried in discrete packets, each unit proportional to the frequency. An even more direct hint came from measurements carried out in 1902 that showed that, once the threshold for bail was met, the liberated electron flew away with an energy proportional to how far the frequency was over the threshold. This was called the photoelectric effect, and Einstein was the only one who correctly interpreted it as signaling a revolution in science. This was one of four papers he wrote in his miracle year of 1905, when he was twenty-six and working in a patent office.

At that time the reigning theory of light was Maxwell's, namely that light is a wave moving through the electric and magnetic fields. Einstein knew Maxwell's theory intimately, having carried Maxwell's book in his pack for a year he spent hiking the mountains as a teenage dropout. No one understood better than Einstein that, great as it was, Maxwell's wave theory of light could not explain the photoelectric effect. For if Maxwell were right, the energy a wave conveys to an electron would increase with intensity, which is exactly what the experiments were not seeing.

The photoelectric effect was not the only clue. The generation of Einstein's teachers had developed the study of light given off by hot bodies, such as the glow of red-hot charcoal. There were beautiful experimental results, which the theorists hoped to explain, which showed that the colors of the emitted light change as the charcoal is heated up. In 1900, theoretical physicist Max Planck explained the result through a derivation that featured one of the most creative misunderstandings in the history of science. To get a glimpse into this comedy, you need to know that even at the turn of the twentieth century, the scientific consensus among physicists, which Planck shared, was that there are no atoms—rather, matter is completely continuous. There were a few prominent theorists who believed in atoms, among them Ludwig Boltzmann of Vienna. Boltzmann developed a method for deriving the properties of gases by treating them as collections of atoms.

Planck, even though he was a skeptic of the atomic hypothesis, borrowed the methods Boltzmann used to study gases and applied them to the properties of light.* Without meaning to do so, he effectively described light as a gas made up of photons, rather than atoms. Navigating in deep waters unfamiliar to him, he found he could get an answer that agreed with experiments if he took the energy of each photon to be proportional to the frequency of the light.

Planck didn't believe in atoms of light any more than he believed in atoms of matter. So he didn't understand that he had made the revolutionary discovery that light is made of particles. But Einstein believed in both, and, almost single-handedly, he understood that the success of Planck's theory rested on treating light as a gas of photons. When he learned about the photoelectric effect, he immediately thought of applying to it the proportionality between the energy of a photon and the frequency of light that had

* For more on how Planck misappropriated Boltzmann's methods, see Thomas Kuhn's *Black-Body Theory and the Quantum Discontinuity, 1894–1912*, or a wonderful biography of Paul Ehrenfest by Martin Klein, both listed in the Further Reading section.

appeared in Planck's work. So it was he, and not Planck, who was given the good fortune of making one of the great discoveries in the history of science: that light has a dual nature, part particle and part wave.

At first Einstein's proposal was greeted with a high degree of skepticism. After all, there was still the double slit experiment to contend with, which clearly showed light traveled through both slits, like a wave. Somehow, light is both wavelike and particle-like. Einstein was to wrestle with this apparent contradiction for the rest of his life. But by 1921 some detailed predictions he'd made in his 1905 paper had been confirmed, and Einstein was awarded the Nobel Prize for the photoelectric effect.

As a footnote to this story, we can mention that another of the four papers Einstein wrote that year gave the final, convincing proof that matter is made of atoms. Atoms were too small to see even with the best microscopes at that time. So Einstein focused his attention on objects just big enough to see through a microscope: pollen grains. These were known to dance unceasingly when suspended in water, which was at the time a great mystery. Einstein explained that the dance was due to the grains colliding with the water molecules, which are themselves constantly moving.*

The other two papers Einstein wrote in that momentous year presented his theory of relativity and the iconic relation between mass and energy: $E=mc^2$.

If we want to find an analogue of what Einstein achieved in that single year, we can only look at Newton. Einstein launched two revolutions—relativity and the quantum. Of the latter he had wrested from nature two precious insights: the dual nature of light,

* Unfortunately, this came too late for Boltzmann, who, depressed at his failure to convince his colleagues of the reality of atoms, committed suicide the next year. And as a footnote to a footnote: a young Viennese physics student called Ludwig Wittgenstein was so dismayed by news of Boltzmann's suicide that he switched to philosophy.

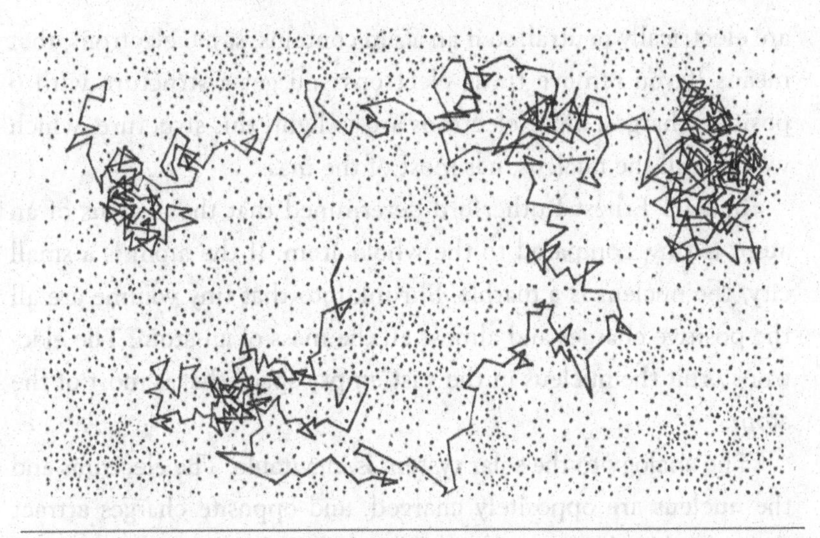

FIGURE 6. BROWNIAN MOTION Brownian motion is the random motion of molecules and other small particles found in nature. Einstein explained that the motion results from the frequent collisions of molecules making up the air or water, and was able to predict how the magnitude of the effects depends on the density of the atoms.

and the relation between the energy of the particle and the frequency of the wave, which ties together the two sides of the duality.

Einstein's fourth paper, which proved the existence of atoms, said nothing about the quantum nature of light. But it contained two mysteries, which it would take the quantum theory to resolve. How could atoms be stable? And why do atoms of the same chemical element behave identically?

While the theorists had been squabbling over whether atoms existed, experimentalists had been busy separating their constituents. First to be identified was the electron, which was revealed to carry a negative charge and to have a tiny mass, about one two-thousandth of that of a hydrogen atom. The chemical elements were understood to be classified by how many electrons they contained. Carbon has 6 electrons, uranium 92, for example. Atoms are electrically neutral, so if an atom contains, say, 6 electrons, that means if you remove those electrons you get a structure with 6 positive charges. Since electrons are so light, this structure, which we can call the nucleus, has most of the mass.

In 1911 Ernest Rutherford determined that the nucleus of an atom is tiny, compared to the whole atom. If the atom is a small city, the nucleus is a marble. Shrunk into that tiny volume are all the positive charges and almost all the mass of an atom. The electrons orbit the nucleus in the vast empty space that is most of the atom.

The analogy to the solar system is inevitable. The electrons and the nucleus are oppositely charged, and opposite charges attract through the electrical force. This holds the electrons in orbit around the nucleus. This much is similar to planets being held in orbit around a star due to their mutual gravitational attraction. But the analogy is misleading because it hides the two puzzles I mentioned. Each provides a reason why Newtonian physics, which explains the solar system, cannot explain atoms.

Electrons are charged particles, and Maxwell's great theory of electromagnetism tells us that a charged particle moving in a circle should give off light continuously. According to Maxwell's theory, which is to say prior to quantum physics, the light given off should have had the frequency of the orbit. But light carries energy away, so the electron should drop closer to the nucleus as its energy decreases. The result should be a quick spiral into the nucleus, accompanied by a flash of light. If Maxwell's theory is right, there can be no picture of electrons circling in gentle, stable orbits around the nucleus. This can be called the crisis of the stability of electron orbits.

You might ask why the same problem doesn't afflict planetary orbits. Planets are electrically neutral, so they don't give off light in the same way. But, according to general relativity, planets in orbit do radiate energy in gravitational waves and spiral into the sun. It is just that gravity is extremely weak, so this process is extraordinarily slow. The effect has been observed in systems consisting of pairs of neutron stars in close orbits. And, very dramatically, gravitational wave antennas have detected the radiation given off by pairs of massive black holes spiraling into each other and merging.

The second problem is why all atoms with a certain number of electrons appear to have identical properties. Two solar systems with six planets each are, beyond that, not generally very similar. The planets will have different orbits and masses and so on. But chemistry works because any two carbon atoms interact with other atoms in exactly the same way. This differs from how oxygen atoms interact, any two of which are also identical to each other. This is the puzzle of the stability of chemical properties. The analogy to the solar system fails because Newtonian physics, which works just fine to explain the solar system, cannot explain why all atoms with six electrons have the same chemical properties.

The answer to both these questions about atoms required applying to atoms the radical new ideas Einstein was developing about the nature of light. This was a bold step of the kind that Einstein was capable of, but even he missed it. The physicist who had the insight was the young Dane Niels Bohr. This insight meant it was Bohr, not Einstein, who would assume the leadership of the revolutionaries who invented quantum mechanics. Throughout his life, Bohr was a radical anti-realist, and it was he, more than anyone else, who was responsible for making the quantum revolution a triumph of anti-realism. Over his career, Bohr fashioned a series of arguments that the behavior of atoms and light could not be understood from a realist perspective.

Bohr grew up in an academic family, the son of a professor of physiology, the brother of a mathematician. He was that fortunate sort who got to live his whole life in the city of his birth, in more or less the same setting as his parents. But in his case, a simple and conservative life was an incubator of radical thought.

In this comfortable, intellectual milieu, he and his wife brought up six sons, several of whom also became professors. One even followed his father to a Nobel Prize in physics. Another son, the oldest, drowned while sailing with his father. Still another son represented Denmark at the Olympics, as did an uncle.

Denmark is a small country that values science, and Bohr's leadership of the quantum revolution was facilitated by the creation of a new institute to support his activities, sponsored by the Danish government and the Carlsberg beer company. This gave Bohr the perfect setting in which to extend his influence, by surrounding himself with the best young theorists from around the world. They were stimulated by a steady stream of visitors who came to collaborate with Bohr or to argue with him about quantum theory. The institute provided him with a comfortable house, where Bohr and his family hosted many of the visitors.

Niels Bohr's sons had to share him with many of these young quantum revolutionaries, who looked up to him as a mentor. His wife looked after them and played matchmaker, introducing several of them to the women who would become their wives. (There were few women who were scientists in Bohr's circle.)

Bohr clearly fascinated those who worked with him. He saw science as a dialogue with nature and his method of working was also based on dialogue—although of a kind that often lapsed into monologue. He used collaborators as scribes, who had the job of taking down Bohr's thoughts, uttered in whispered riddles, corrected and corrected again, as Bohr paced in circles around the room.

Bohr began to work on quantum physics shortly after receiving his PhD. He went right to the heart of the problem by proposing a simple but radical quantum model of the atom. He built on Einstein's nascent quantum theory, particularly the idea that energy is carried by photons. To address the problem of the stability of the electron orbits, Bohr simply postulated that Maxwell's theory is wrong on the atomic scale. He hypothesized, instead, that there are a small number of orbits of the electron, which are stable. To distinguish these good orbits, he made use of Planck's constant, which is the conversion factor between frequency and energy. This conversion factor has units of a quantity called angular momentum. This works just like momentum, but for circular motion. A spinning body has an inertia to continue rotating. This is because spinning or orbiting bodies carry angular momentum, which, like energy and regular momentum, cannot be created or destroyed. It is this conservation of angular momentum that keeps a bicycle wheel spinning; it is also what causes a figure skater to spin more rapidly when she pulls her arms in.

Let's think about a hydrogen atom, which has only a single electron. Bohr postulated that the good orbits are those in which the electron has certain special values of angular momentum. These special values are integer multiples of the unit of angular momentum, given by Planck's constant. Bohr called these *stationary states*. There is an orbit with zero angular momentum which also has the lowest possible value of energy for an electron in orbit around the nucleus. This state is stable; it is the ground state. At higher energies above the ground state are a discrete series of energies which are the excited states.

Atoms can absorb light, gaining energy, and they can also radiate energy away by giving off light. Bohr next postulated that these processes happen when the electron jumps between the stationary states. To describe these jumps, Bohr made use of Einstein's photon hypothesis. When an electron jumps down from an excited state to the ground state, it gives off a photon. That photon has an energy equal to the difference in energies of the two states, so that the total energy is unchanged. It has a specific frequency, given by Planck and Einstein's relation between frequency and energy.

If you reverse this process you can cause an electron to jump from the ground state up to an excited state, by giving it a photon with an energy equal to the difference of the two states.

A given atom can then give up or absorb light only at the special frequencies that correspond to these energy differences between states of its electrons. These special frequencies are called the spectrum of the atom.

By the time Bohr worked this all out, in 1912, chemists had measured the spectrum of hydrogen. Using the ideas I've just described, Bohr was able to calculate the spectrum, and his simple theory reproduced what the experimentalists had seen.

This was a huge step, but it was only a first step toward an understanding of the quantum. There remained many open questions and problems. What is an electron such that it can travel freely outside the atom, but can exist only in one of the stationary states when in an atom? And, most urgently, can the theory be applied to atoms besides hydrogen?

The next decade was taken up by numerous clever attempts to apply Bohr's theory to different atoms and other systems. We can generously say the results were mixed, even as we admire the ingenuity of the attempts. This was the situation by the time a young French aristocrat named Louis de Broglie started graduate school in Paris around 1920.

Louis Victor Pierre Raymond, duc de Broglie, was born of a noble family in the last years of the nineteenth century and studied history before switching to physics. He served in the army during the First World War in the wireless telegraphy section; he was stationed at the Eiffel Tower.

The small world of theoretical physics was then, as it is now, intensely social. During the crucial period when quantum mechanics was being developed, the proponents were continually in touch by letter and postcard, and they made frequent train trips to visit and consult. The aristocrat de Broglie was an outsider to this world by dint of his personality and position, and because Paris was at the time a backwater in theoretical physics. Louis de Broglie spoke regularly about his work with only one person, his brother, Maurice de Broglie, an experimental physicist who worked on X-rays.

Isolation is usually an obstacle for scientists, but sometimes it can lead to someone stumbling on an insight that everyone in the crowd has missed. De Broglie was still a doctoral student when he shook physics to the core by putting forth an audacious hypothesis: that the wave-particle duality is not just a feature of light—it is universal. In particular, electrons, like light, are waves as well as particles.

As he remarked, "When in 1920 I resumed my studies . . . what attracted me . . . to theoretical physics was . . . the mystery in which the structure of matter and of radiation was becoming more and more enveloped as the strange concept of the quantum, introduced by Planck in 1900 in his researches into black-body radiation, daily penetrated further into the whole of physics."[1]

The power of a fresh mind taking a fresh look at a problem is one of the wonders of the world. The young de Broglie had the obvious idea, which had somehow eluded even Einstein and Bohr. They sought to avoid the embarrassment of the wave-particle duality. De Broglie doubled down on it. If light was both a wave and a particle, why couldn't the same be true of electrons? Why not hypothesize that the wave-particle duality applies universally to all matter and radiation?

As de Broglie later recounted it, "As in my conversations with my brother we always arrived at the conclusion that in the case of X-rays one had both waves and corpuscles, thus suddenly . . . I got the idea that one had to extend this duality to material particles, especially to electrons."[2]

What motivated de Broglie to come up with an idea which many more experienced physicists had missed? De Broglie was engaged in an ambitious project to reinvent physics from the ground up to incorporate the wave-particle duality. He started with light, where there was already good evidence for a duality of waves and particles, and asked a simple question few had asked before: How do the light quanta move?

Recall that Newton had favored a particle theory of light because he believed that particles travel in straight lines. The same assumption had led Thomas Young to abandon the particle picture and embrace the idea that light is a wave when he understood that light could bend when diffracted by an obstacle or refracted by passing between two media. It makes sense that if light doesn't travel in straight lines, it is not made of particles. What then of photons? Didn't they have to travel in straight lines? De Broglie's idea was that they don't because they are guided by the waves, which do diffract and refract.

This is stunningly revolutionary. The idea that particles travel in straight lines is a consequence of the most basic principle in all of physics, which is Newton's first law of motion. Also called the principle of inertia, it states that a particle with no forces on it moves at a constant speed in a straight line. One consequence is that momentum is conserved. It is also closely related to the principle of relativity, for another consequence is that velocity is a purely relative quantity.

De Broglie understood that light quanta were going to have to bend around obstacles, violating all these fundamental principles. The goal of his thesis was to formulate a revolutionary new theory of motion, which would apply to the particles contemplated by the wave-particle duality. In this context, it was a small and necessary step to extend the wave-particle duality from light to all forms of matter and energy.

In 1924 he wrote this up as his PhD thesis. The thesis was short and uncompromising. The legend is told that had he not been from the aristocracy, it is possible de Broglie would simply have been failed. Not knowing what else to do, his committee sent the thesis to Einstein to evaluate. Einstein saw de Broglie's point and recommended approval. At the same time, he sent de Broglie's thesis to a few people he knew would be very interested in it.

One of these was his friend Max Born, then a young professor in Germany. An experimentalist colleague of his, Walter Elsasser, heard of it and suggested that de Broglie's prediction that electrons could be diffracted might be tested by scattering a beam of electrons off a crystal. Max Born passed the suggestion to experimentalists in England. None succeeded, but meanwhile two American experimentalists working at Bell Labs, Clinton Davisson and Lester Germer, were, for other reasons, studying how electrons scatter off the surfaces of metals. They accidentally discovered the diffraction of electrons when, in 1925, they tried a new procedure which had the unintended consequence of developing a layer of atoms organized in the regular arrays of a crystal on the surface of their sample. When they measured where the electrons went that scattered off the metal with the crystal surface, they saw interference patterns. Davisson was unaware of the significance of this until he attended a conference in Oxford in the summer of 1926, and happened to listen to a talk by Max Born, who showed a figure from one of Davisson's own papers as evidence for de Broglie's revolutionary hypothesis of matter waves. When Davisson returned, he and Germer went back to the lab and were able to definitively confirm that electrons diffract, just as de Broglie had predicted.

Erwin Schrödinger was a brilliant mathematical physicist, originally from Vienna, who had become a professor at the University of Zurich. Schrödinger was closing in on forty and did not belong to the young generation of de Broglie and the other physicists who were revolutionizing their field. On November 23, 1925, he attended a colloquium by Peter Debye, who gave an enthusiastic presentation of de Broglie's matter wave hypothesis. Debye ended by saying there was one thing missing from de Broglie's beautiful picture: an equation to describe how the electron waves travel in space. Leaving his wife behind in Zurich, Schrödinger took de Broglie's papers with him to a Christmas holiday in the mountains with his girlfriend. (His wife was spending the Christmas holidays with her lover, the great mathematician Hermann Weyl, who was also Schrödinger's best friend.) The first day, he excused himself from skiing, stayed in their chalet room, and read de Broglie's papers. He challenged himself to invent the equation that would govern de Broglie's electron wave. He succeeded the next day, and by the time he returned from the mountains, he had captured the equation that bears his name, the fundamental equation of quantum theory.

Not only that, but, shortly after returning, with the help of Weyl, Schrödinger solved his equation for the case of a single electron in orbit around a nucleus, and reproduced Bohr's theory of stationary states and his prediction of the spectrum of hydrogen. The key idea is that the electron waves have to fit around an orbit, as we see in figure 7. The thoughts of the girlfriend—and, indeed, her name—are lost to history. But legend tells us that when Schrödinger went to Stockholm to receive his Nobel Prize he showed up with his wife and their girlfriend.

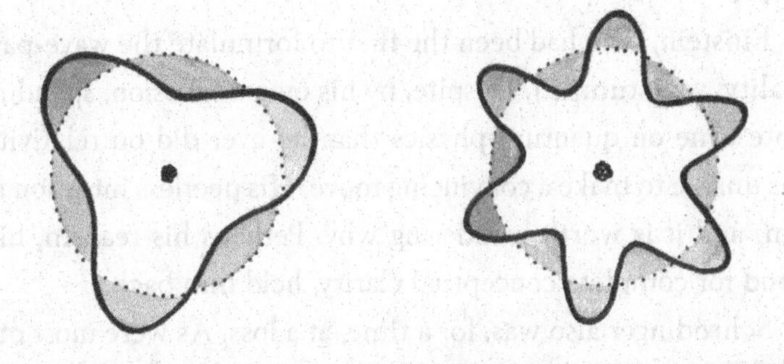

FIGURE 7. Electron waves in the atom. The wave on the left fits around the nucleus in three steps, so the wavelength is the diameter of the atom divided by three. The right figure has half the wavelength and so fits around in six steps.

Thus quantum mechanics was born. The question everyone then faced was how to think of the electron wave that de Broglie had invented and Schrödinger had tamed. Schrödinger at first thought that the electron simply is a wave. This didn't hold up because it was easy to show that the wave tended to spread out in space as it traveled, whereas one could always find a localized particle. Max Born then proposed his rule that the wave is related to the probability of finding the particle.

For Einstein, the wave-particle duality, while a profound challenge, had been limited to speculation about the constitution of light. Confined to that domain, it did limited damage, perhaps because particle and wave theories of light each had long histories and recognized virtues. But the idea of matter waves came as a complete shock. De Broglie and Schrödinger transformed physics by bringing the wave-particle duality into the core of physics, where it sat enshrined as the central mystery of the revolutionary new quantum physics.

The question was no longer "How can light be both a particle and a wave?" but rather, "How can *everything* be both a particle and a wave?"

Einstein, who had been the first to formulate the wave-particle duality, was stumped. Despite, by his own admission, spending far more time on quantum physics than he ever did on relativity, he was unable to make a convincing move. His peerless intuition failed him, and it is worth wondering why. Perhaps his realism, his demand for complete conceptual clarity, held him back.

Schrödinger also was, for a time, at a loss. As were most others.

Of the great pioneers, only Bohr knew what to do. It was his moment and he seized it, announcing the birth not just of a new physics but of a new philosophy. The moment for radical anti-realism had come, and Bohr was ready for it.

Bohr called the new philosophy *complementarity*. Here is how he talked about it: Neither particles nor waves are attributes of nature. They are no more than ideas in our minds, which we impose on the natural world. They are useful as intuitive pictures that we construct from observing large-scale objects such as marbles and water waves. Electrons are neither. Electrons are microscopic entities that we cannot observe directly, and so we have no intuition about them. To study electrons we must construct big experimental devices to interact with them. What we observe is never the electron itself; it is only the responses of our big experimental devices to the tiny, invisible electrons.

To describe how the experimental devices respond to electrons, we may find it useful to employ intuitive pictures such as the wave picture or the particle picture. But we cannot take these pictures too seriously because different experiments require different pictures. The different pictures would contradict each other if we forgot the context and applied them to the electrons themselves. But there is no actual contradiction so long as we remember two things. The pictures are useful only as a description of an electron in a specific context, which is in a particular experimental device. And there is no experimental device that forces us to apply both contradictory pictures simultaneously.

Bohr's position is anti-realist in the extreme, in that he denies it is even possible to talk about or describe an electron as it is in itself, outside the context of an experiment we construct. Science according to this picture is not about electrons; it is about how we talk about our interactions with them.

For Niels Bohr, complementarity was more than a principle; it was a proposal for a whole philosophy of science. And what a radical proposal it was. Bohr championed the philosophy of complementarity throughout his life, as did other founders of quantum mechanics, including, to some extent, Heisenberg.

For Bohr, science is not about nature. It does not and cannot give us an objective picture of what nature is like. That would be impossible, because we never interact with nature directly. We gain knowledge about the natural world only through intermediaries, which are experimental devices we invent and construct.

Thus, we must give up the idea that science gives us an objective description of nature, or has anything at all to say about what nature is like, absent our existence and our interventions. Science is rather an extension of a common language we use to describe to each other the results of our interventions into nature.

In essays and books, Niels Bohr argued that his philosophy of complementarity had wide applicability. It has been claimed he got the idea of complementarity from the Kabbalah, the Jewish mystical writings, which speak of the complementarity between God's love and God's justice. Bohr talked about the complementarity between life and physics, between energy and causation, and, indeed, between knowledge and wisdom. For Bohr the lesson of quantum mechanics was a revolution that extended beyond physics, beyond science.

ONE REASON QUANTUM MECHANICS captured the interest of the younger generation of physicists was that it could be approached from several points of view. I have so far told the story of one way the quantum theory was invented, centering on the wave-particle duality, but there was another route, which had been discovered shortly before Schrödinger took his Christmas holiday. This was pioneered by Werner Heisenberg, a young and very confident German theorist, who completed his education in Max Born's group in Göttingen and then in 1925 went on a research fellowship to work in Copenhagen with Bohr. He spent the next several years bouncing between Göttingen and Copenhagen, which is to say he was in close touch with the two most dynamic scientific personalities of that moment, Born and Bohr. Max Born and several of his students and assistants also played important roles in the story; indeed, the full story of how quantum mechanics was invented involves at least half a dozen theorists, in frequent communication.

Heisenberg worked from a particular idea about physics, an idea that was anti-realist to begin with. He asserted that physics does not give a description of what exists, as realists suppose, but is only a way to keep track of what is observable. For large-scale objects, we have gotten used to confusing the two. But if we want to make sense of atomic physics, we must adhere strictly to the dictum that science can only refer to what can be observed.

Hence, Heisenberg asserts that it is meaningless to talk about how the electron moves in the atom, unless that motion has consequences which can affect large-scale measuring devices. According to Bohr's model, an atomic electron spends most of its time in stationary states, during which it has no interaction with anything outside the atom. It is then meaningless to ask how the electron moves while it is in a stationary state. It is only when it jumps between stationary states that the atom can interact with the world outside, because the jump is accompanied by the absorption or creation of a photon, and that photon's energy can be measured by a spectrograph.

Heisenberg's admonition not to try to model the trajectories of electrons in stationary states must have come as a breath of fresh air to others of his generation who were spending much of their time in frustrating and ultimately fruitless attempts to do just that.

Heisenberg was inspired by this thinking to invent a new way of representing the energy of the electron. Not by a single number, because to do so would be to claim that the energy is a property of the atom alone. What is relevant for physics is only what aspect of energy affects a measuring device. These are the energies carried by the photons that the atoms absorb or emit when the electrons jump between energy levels. These are the differences between the energies in the different stationary states.

Heisenberg arranged these energy differences as a table of numbers. He then imagined that such tables could represent observable aspects of other quantities, such as the electron's position and momentum. To make a theory he had to do more, which was to find a way to write equations involving these tables of numbers. In the equations of physics we often find ourselves adding or multiplying numbers. He needed to do the same with tables of numbers. So he had to invent rules for how to do this.

As a member of both Bohr's institute and Max Born's research group, Heisenberg was under the influence of two masters with very different styles of work, and the contrast between them undoubtedly stimulated his thinking. But to realize his ideas in detail, he needed isolation, no less than Einstein, de Broglie, and Schrödinger had. Like Schrödinger, he took off on a holiday, in his case to a small island called Helgoland.

Once there, it took him only a few days to take himself on the journey I've just sketched, and to invent ways to write and solve equations with his tables of observable quantities.

He tested his ideas on a simple toy model of an atom, in which the electron is bound by a constantly increasing force, as if on a spring. This was not meant to be realistic, but it was a simple test, because the answer was known, and his method passed. There was only one hitch: he discovered that the order in which he multiplied two tables together matters. In the language I proposed earlier, Heisenberg's tables of numbers don't commute. This is of course not the case for ordinary numbers, and at first this discovery dismayed Heisenberg.

Nonetheless, he wrote up his findings in a paper published at the end of 1925. It was in the introduction to that paper that he announced his program of constructing laws of physics that dispensed with mechanical models describing the trajectories of the electrons and involved only relationships between observable quantities, namely the spectra of light the atoms emit and absorb.

This was a big step, but it was not yet the complete theory. He then returned to Göttingen and worked with Max Born and a brilliant student of his, Pascual Jordan. Born and Jordan were already partway to a new theory, and explained to Heisenberg that his tables of numbers were known to mathematicians as matrices; and they were able to reassure him that the failure to commute was a feature and not a bug. Heisenberg then understood that since the tables/matrices represent a process of measurement, the order does matter—because it matters in which order we make measurements. Together the three theorists then worked out the rest of the new theory, which they named quantum mechanics. A joint paper by the three of them was the first complete statement of the new theory.

Austrian wunderkind Wolfgang Pauli quickly followed up and applied the new theory to find the spectrum of the hydrogen atom, and it came out exactly right. Thus was quantum mechanics born by a second route, and in a way that was directly inspired by the anti-realist principles Heisenberg had expressed in his 1925 paper. The new theory of Born, Heisenberg, and Jordan is expressed in terms of quantities that describe how an atom responds to being probed by an external measurement device; there are no quantities that describe the exact trajectories of the electrons, independent of our interactions with them.

One quantum theory of the atom is great, but two are a problem, especially since they both reproduced the right spectrum of hydrogen. The two theories could not have differed more, as reflects the philosophies of their discoverers. Einstein, de Broglie, and Schrödinger were realists. Even if there were mysteries, they believed an electron was real and somehow existed as both wave and particle. Bohr and Heisenberg were enthusiastic anti-realists who believed we have no access to reality, only to tables of numbers which represent the interactions with the atom, but not the atom directly.

The tension lasted a few months, and then had an unexpected resolution when Schrödinger showed that the two forms of quantum mechanics are completely equivalent. Like two languages, you could speak in terms of waves or talk the language of matrices, but the math problems you had to solve turned out to be just different expressions of the same logic.

Heisenberg and Bohr, together in Copenhagen, shared an anti-realist perspective. They sought a way to speak consistently about properties that could not be realized simultaneously, such as waves versus particles or position versus momentum. Bohr's resolution of the apparent paradoxes was his principle of complementarity. Heisenberg's was his great uncertainty principle, which we talked about in chapter 2.

The uncertainty principle is a very general principle, as it says that we cannot know exactly both where a particle is and with what momentum it is moving. It has, as Heisenberg and his mentor Bohr realized immediately, stunning consequences. One is that the determinism of Newtonian physics cannot survive in the quantum world, because to predict the future motion of a particle you must know both its present position and how fast and in what direction it is moving, and hence its momentum. If you cannot know both precisely, you cannot predict where the particle will be at later times. As a result, the best that quantum theory can do is to make probabilistic predictions about the future.

The consistency of complementarity depends on there never being a case where we are forced to use both the particle picture and the wave picture in the description of a single experiment. The impossibility of doing so is safeguarded by Heisenberg's uncertainty principle, which he proposed in 1927, after he had moved back to Copenhagen and was in close contact with Bohr.

Historians tell us that luck plays a big role in science. Heisenberg was doubly fortunate for, as the protégé of both Max Born and Niels Bohr, he was not just in the right place at the right time, but doubly so! From his mentor Bohr he was inspired to abandon realism and model the atom only in terms of the energies it exchanges with our measuring devices, and from his mentor Born he got the mathematical tools needed to give these ideas a precise expression.

Of course, Heisenberg knew his good fortune and was the one who pushed to frame the new theory precisely. There were perhaps half a dozen young theorists who were also in the orbits of Bohr and Born, who contributed pieces, like Pauli, or got partway there, like Jordan, or were a few months late and so got to elegantly frame the new theory, like the English theorist Paul Dirac. The full story of the invention of the matrix form of quantum mechanics is far more complex than I can tell here, as it reveals a very dynamic, collective effort of a diverse community of theorists, in close interaction.

Still, diverse as they were, the matrix mechanicians were by 1927 all framing the new theory in terms of the radically anti-realist philosophy that Bohr preached. The only holdouts were those who had come to quantum mechanics through the wave-particle duality, Einstein, de Broglie, and Schrödinger, who stubbornly remained realists. But once it was proved that Schrödinger's wave mechanics was equivalent to Heisenberg's matrix mechanics, the realists could be dismissed as stubbornly grasping on to old metaphysical fantasies, and ignored.

The essence of Bohr's philosophy is the necessity of basing science on incompatible pictures and languages. Heisenberg preached a view which differed in emphasis from Bohr's while being loosely compatible with it. Heisenberg emphasized that science concerns only measurable quantities and can't give an intuitive picture of what is happening at atomic scales. The observable quantities relevant for interacting with an atom include the energies and lifetimes of the stationary states, but do not include the positions or motions of electrons in their orbits around the nucleus. So quantum physics only has to yield an answer to a question of where an electron is if you force it into a context where that position is measured. According to Heisenberg, observable quantities are brought into existence only by the act of measuring them. When an atom is free of a measuring apparatus, no quantity describes it.

This may be called an *operationalist perspective*. It is certainly anti-realist, in that Heisenberg stressed that this view is mandatory. There was, according to him, no possibility of seeing deeper into the atom to perceive how the electrons move in their orbits. His uncertainty principle precluded it.

Heisenberg explained that uncertainty and complementarity were closely connected.

> We can no longer speak of the behavior of the particle independently of the process of observation. As a final consequence, the natural laws formulated mathematically in quantum theory no longer deal with the elementary particles themselves but with our knowledge of them. Nor is it any longer possible to ask whether or not these particles exist in space and time objectively. . . .
>
> When we speak of the picture of nature in the exact science of our age, we do not mean a picture of nature so much as a *picture of our relationships with nature*. . . . Science no longer confronts nature as an objective observer, but sees itself as an actor in this interplay between man and nature. The scientific method of analyzing, explaining and classifying has become conscious of its limitations, which arise out of the fact that by its intervention science alters and refashions the object of

> investigation. In other words, method and object can no longer be separated. . . .
>
> [T]he different intuitive pictures which we use to describe atomic systems, although fully adequate for given experiments, are nevertheless mutually exclusive. Thus, for instance, the Bohr atom can be described as a small-scale planetary system, having a central atomic nucleus about which the external electrons revolve. For other experiments, however, it might be more convenient to imagine that the atomic nucleus is surrounded by a system of stationary waves whose frequency is characteristic of the radiation emanating from the atom. Finally, we can consider the atom chemically. . . . Each picture is legitimate when used in the right place, but the different pictures are contradictory and therefore we call them mutually complementary.[3]

Bohr's point was even more radical. For him,

> An independent reality in the ordinary physical sense can . . . neither be ascribed to the phenomena nor to the agencies of observation. . . .
>
> A complete elucidation of one and the same object may require diverse points of view which defy a unique description. Indeed, strictly speaking, the conscious analysis of any concept stands in a relation of exclusion to its immediate application.[4]

Other quantum luminaries, such as Wolfgang Pauli, a wunderkind who published a textbook on general relativity when he was twenty-one, and John von Neumann, a Hungarian mathematician who is famous for his inventions in a broad range of fields, from the architecture of computers to the mathematics of quantum theory, taught variants of these anti-realist philosophies. Their views differed in emphasis, but anything written by them was classified as part of the "Copenhagen interpretation" of quantum mechanics. This name recognized Bohr's dominance as the oldest of the group and mentor to most, as well as the originator of nothing less than a new way of talking about science. The name also recognized Bohr's institute as the central node in the network of quantum physicists, where they all studied, worked, or visited.

One of the hardest lessons to learn in academic life—and for me one of the most disconcerting—is the speed with which a radical insurgency can become orthodoxy. In just a few years a generation of students championing a dangerous new idea are elevated by an initial success into professorships. From these positions of influence they form a powerful network of academic power brokers, which they use to ensure the continuation of the revolution. Such was the case with the generation of quantum revolutionaries. In 1920 Heisenberg was a student, as were Dirac, Pauli, and Jordan; 1925 found them young researchers fully engaged in the invention of quantum theory; by 1930 they were senior professors, and the revolution was over. The fact that there remained a handful of defectors—Einstein and Schrödinger from the older generation, and de Broglie among their contemporaries—did nothing to diminish their triumph, for students knew which way the wind blew and followed the ascendant orthodoxy. For the next half century, the anti-realism of the Copenhagenists would be the only version of quantum theory taught.

PART 2

REALISM REBORN

SEVEN

The Challenge of Realism: de Broglie and Einstein

There was never a single Copenhagen interpretation. Bohr, Heisenberg, and von Neumann each told a somewhat different story. But they all agreed that science had crossed a threshold. There could be no retreat back to a realist version of physics. They gave diverse arguments against the possibility, all leading to the conclusion that quantum physics is inconsistent with realism. No version of atomic physics could exist if it included electrons with definite positions and trajectories.

One way all these arguments might have been defeated—one would think—was for someone to come up with an alternative quantum theory based on realist ideas.

What is really bizarre, looking back, is that from 1927 on, there had existed a realist version of quantum mechanics. This is based on a stunningly simple idea. Perhaps you have already thought of it. It is simply to posit that there are *both* waves *and* particles. What gets created and detected, what gets counted, is a particle.

Meanwhile, a wave flows through the experiment. The wave *guides* the particle. The result of this guidance is that the particle goes to where the wave is high.

Faced with a choice of which way to go around an obstacle, such as in the double slit experiment, the wave goes both ways. The particle goes through only one slit around only one side, but where it goes once it gets through is guided by the wave, and shows the influence of both paths.

This obvious solution to the challenge of the wave-particle duality was thought up by Louis de Broglie. He worked it out in detail and called it the *pilot wave theory*. De Broglie presented his theory at a famous conference held in Brussels in 1927. Named the Fifth Solvay Conference after its sponsor, the conference featured talks by most of the revolutionaries of the new quantum physics.

The core of pilot wave theory was de Broglie's idea that the electron is actually two entities, one particle-like and one wave-like. The particle is always located some particular place and always follows some particular path. Meanwhile, the wave flows through space, taking simultaneously all the possible paths or routes through the experiment. The wave then directs the particle where to go, and that piloting will be based on conditions along all the paths. Even though the particle must take one route or another, which route it takes is influenced by the wave, which flows through all routes.

This influence of a wave on a particle is the new thing which is responsible for much of what is strange in the quantum world. There are two laws, one for the wave and one for the particle. The wave law is relatively familiar; it is not so different from the laws that physicists use to describe sound waves or light waves. The waves spread out, and as they travel they diffract and interfere. Like water and sound, these quantum waves will flow down every channel open to them. And when waves coming down different channels meet, they will interfere.

The wave in question is called the wave function. It propagates according to the simple equation that Schrödinger invented during his romantic ski weekend. This is Rule 1, and it is the key equation in every approach to quantum physics.

There is no Rule 2 in this framework. But there is a new law that directs the particle to follow the wave, which is called the guidance equation. The system defined by the wave function together with the particle evolves deterministically, which suggests it is complete.

In other approaches to quantum mechanics, it is simply posited that the particle will be found where the wave is large. More precisely, the probability of finding the particle at some particular place is proportional to the square of the wave function there. This is what we earlier called the Born rule.

In pilot wave theory it is also the case that the particle is more likely to be found where the wave is high. But this is not posited. Rather, it turns out to be a consequence of the law that drives the particle to follow the wave.

Place a ball on a hillside and watch it roll down from there. You may observe that the ball tends to follow the steepest path downward. This is called the law of steepest descent. Roughly speaking, de Broglie's guidance equation does the opposite, guiding the particle on the steepest path to climb the wave function.* We can call it the law of steepest ascent. A mountain climber following this law would at each moment of her climb choose to go in the direction of the steepest slope of the mountain.

De Broglie was able to demonstrate that the probability law

* I am oversimplifying a bit. The particle follows a part of the wave function called its phase.

posited by Max Born is a consequence of the particle following the steepest ascent. To illustrate this important point, imagine that you throw a bunch of particles down on a hillside representing the wave function. Wherever you throw them, the particles will quickly arrange themselves so that they are more likely to be found where the square of the wave function is largest, which reproduces Born's law.

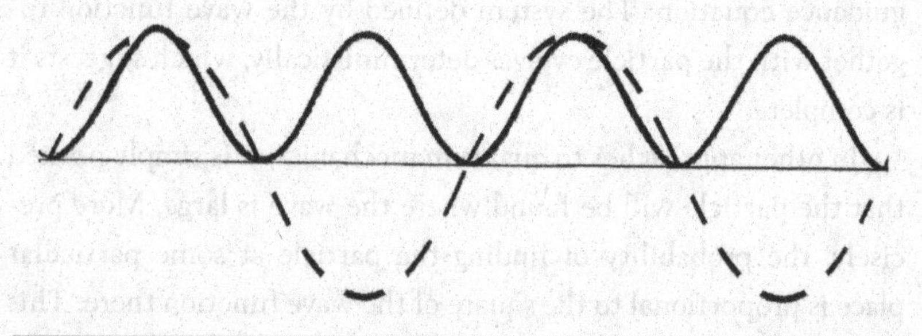

FIGURE 8. SQUARING THE WAVE FUNCTION The dashed line denotes a wave traveling along the horizontal line to the right. Note that it spends as much time with negative values as it does with positive. The solid line is the square of the wave, which is always above zero.

The pilot wave theory predicts everything quantum mechanics does, but explains a good deal more. The mysterious way in which the ensemble seems to influence the individual is cleared up and explained straightforwardly as the influence of the wave on the particle. Both are real, and both exist for every individual atom. Everything that was puzzling and mysterious about quantum mechanics is revealed to be a consequence of that theory leaving out half of every story.

Despite what Bohr and Heisenberg say, the electron always has a position and it follows a definite trajectory, which is perfectly predictable if you know the right law. No need for operationalism, and no sense wasting time trying to make sense of Bohr's obscure pronouncements on complementarity. Waves and particles don't contradict each other; instead, both are always present and they work together to explain atomic physics. What is, simply is.

There is an alternative history in which all the bright, ambitious students flocked to Paris in the 1930s to follow de Broglie, and wrote textbooks on pilot wave theory, while Bohr became a footnote, disparaged for the obscurity of his unnecessary philosophy. It was, alas, not to be. But why the convoluted philosophy of complementarity triumphed, while it was de Broglie's pilot wave theory which became the forgotten footnote, is a question to be pondered.

The pilot wave theory overlaps with quantum mechanics, but it also differs on several points. Rule 1 is common to quantum mechanics and pilot wave theory. But pilot wave theory differs from quantum mechanics in having no Rule 2. Instead there is a law to guide the particle. The laws of pilot wave theory are deterministic.

With no Rule 2, the quantum state in pilot wave theory never collapses. This has some strange consequences which took its adherents some time to appreciate, and to which we will return in the next chapter.

AT THE SOLVAY CONFERENCE the talks were followed by discussions, and these were transcribed and published in a book with the talks. There is not much evidence that de Broglie's presentation changed minds, although it was discussed. One person who did get it and did comment was Einstein.

Although he doesn't say so in the transcribed discussion, Einstein had himself thought of the idea of pilot wave theory. In May 1927 Einstein gave a talk to the Prussian Academy of Sciences in which he presented a rather complicated version of the pilot wave idea. He discussed the idea in correspondence with Heisenberg and others and submitted a paper based on the talk for publication. But just before it was to appear, Einstein withdrew his paper.* He had apparently realized his version of pilot wave theory had several problems, some of which prevented the theory from reproducing all the predictions of quantum mechanics. So far as is known, he never mentioned it again.

Einstein had been scheduled to give a talk at the Solvay conference, probably about that paper. He backed out of that talk at the last minute, writing to the conference organizer, "I kept hoping to be able to contribute something of value in Brussels; I have now given up that hope. . . . I did not take this lightly but tried with all my strength."[1]

Einstein nonetheless did attend, and, of course, he contributed to the discussions that took place about the new quantum theory. Among them were the first discussions between him and Bohr in which Einstein tried to find inconsistencies in the new quantum mechanics. These intense discussions were informal and unfortunately were not transcribed. But much later, Bohr published his reminiscences of those discussions, in a paper that is both one of the most compelling reads in the history of physics and a masterpiece in scholarly propaganda.

During the meals and breaks of the conference, Einstein presented Bohr with several arguments that quantum mechanics is inconsistent. He posited that to give a complete description would require additional variables, which are hidden in the quantum mechanical description. Bohr doesn't mention that this is what de Broglie had achieved with his pilot wave theory. On the contrary, in

* Antony Valentini gave me a copy of that paper of Einstein's during a speech at my wedding, which I promptly lost.

Bohr's telling he was able, after a sleepless night, to refute Einstein's objection, leaving in place his view as to the consistency and even the inevitability of complementarity.

Einstein responded positively during the discussion of de Broglie's talk. After describing an objection to the Copenhagen version, he said, "In my opinion, one can remove this objection only in the following way, that one does not describe this process solely by the Schrödinger wave, but that at the same time one localizes the particle during the propagation. I think that Mr. de Broglie is right to search in this direction."[2]

Few quantum physicists mentioned de Broglie's theory in the years after its presentation in 1927. In spite of de Broglie being justly admired for his insight of extending the wave-particle duality to matter, and in spite of his having presented the pilot wave theory at the most important conference on quantum physics, with an audience of virtually everyone who mattered in atomic physics, it was as if de Broglie had never published or presented his theory. So far as I know, no textbooks mentioned it for decades after. It is not that there were Copenhagen textbooks and pilot wave textbooks. There were only Copenhagen textbooks. These either ignored the foundational issues with the theory or presented a confident assertion that all questions that were meaningful had already been answered by Bohr and Heisenberg.

One important reason anti-realism triumphed was that the mathematician John von Neumann published a proof he claimed showed there could not be a consistent alternative to quantum mechanics. This was published a few years after the Solvay conference in a book on the mathematical structure of quantum mechanics. This claim had to be wrong, as it implied de Broglie's pilot wave theory had to be inconsistent, which it wasn't. You might have thought that someone would have mentioned this.

Von Neumann's incorrect proof seems to have been one of those cases that happens far too often in the history of science, where a result is as influential as it is wrong. Von Neumann had a formidable reputation, and in the face of his theorem, opposition to the view that quantum mechanics was the most complete theory possible caved. In particular, de Broglie himself capitulated to the combined criticisms of von Neumann and other theorists, including Wolfgang Pauli.

It is not quite true that nobody noticed that von Neumann's theorem contained a mistake. A young mathematician called Grete Hermann took an interest in quantum mechanics and was naturally drawn to study von Neumann's book. A good mathematician in her own right, Hermann was a PhD student of Emmy Noether,* and among her accomplishments are several results which anticipated the modern study of algorithms in computer science. She also had a keen interest in philosophy and was concerned with the implications of quantum mechanics for the neo-Kantian philosophy then popular in the German-speaking academy.

Grete Hermann quickly noticed a mistake in the proof of the theorem on the impossibility of hidden variables in von Neumann's book. One of the assumptions of the theorem was already equivalent to the basic structure of quantum mechanics. So all the theorem proved was that any theory equivalent to quantum mechanics would turn out to be equivalent to quantum mechanics.

Very unfortunately, the paper she wrote exposing the fault in von Neumann's proof had no impact.[3] Part of the reason may have been that she published it in an obscure journal, but it is hard to

* Noether is one of the greatest twentieth-century mathematicians; among her many discoveries was a seminal theorem on symmetry in physics, which we will come to.

avoid the thought that she wasn't taken as seriously as she might have been due to her gender, as well as to the fact that her paper punctured one of the main arguments used to establish the inevitability of quantum mechanics.

It took two long decades for someone else to notice that von Neumann's proof had to be wrong, because it disagreed with the manifest existence of pilot wave theory. This was David Bohm, who will be the protagonist of the next chapter. Ten years after that, John Bell isolated the error as an erroneous assumption. Here is how he put it:

> [T]he von Neumann proof, if you actually come to grips with it, falls apart in your hands! There is *nothing* to it. It's not just flawed, it's *silly*. . . . When you translate [his assumptions] into terms of physical disposition, they're nonsense. You may quote me on that: The proof of von Neumann is not merely false but *foolish*![4]

David Mermin, in a lucid review of various impossibility theorems, regretted the "many generations of graduate students who might have been tempted to try to construct hidden variables theories [who] were beaten into submission by the claim that von Neumann . . . had proved that it could not be done." Mermin "wonder[ed] whether the proof was ever studied by either the students or those who appealed to it to rescue them from speculative adventures."[5]

It is hard now, looking back from our present vantage point, in which several competing views about how to understand quantum theory flourish, to appreciate the state of mind of the first several generations of quantum physicists. In spite of the persistent and powerful dissents of Einstein, de Broglie, and Schrödinger, for at least the first half century following the invention of quantum mechanics in 1925, the anti-realist philosophy initiated by Bohr dominated all discussions of quantum theory.

Through all those years, if someone raised the possibility of a realist version of quantum mechanics, the response, I was told, was a good dose of Copenhagen-speak which, if one persisted, was capped off with "Von Neumann proved there is no alternative." One can imagine it would have changed things at least a little if Grete Hermann's paper showing that no, von Neumann hadn't proved anything, had been known. But it simply wasn't.

EIGHT

Bohm: Realism Tries Again

In 1952, David Bohm solved the biggest of all problems in quantum mechanics, which is to provide an explanation of quantum mechanics. . . . Unfortunately, it is widely under-appreciated. It achieves something that was often (before and even after 1952) claimed impossible: To explain the rules of quantum mechanics through a coherent picture of microscopic reality.

—RODERICH TUMULKA

By 1930 de Broglie had given up. From then on, the anti-realist Copenhagen interpretation dominated the teaching and application of quantum mechanics, as well as most discussion of the new theory's implications. The only significant exceptions were Einstein and Schrödinger, who continued to challenge the Copenhagen school and insist on the need for a realist formulation of quantum theory. But their dissent had little impact.

That was the situation in the early 1950s when the young American theorist David Bohm set out to write a textbook on quantum mechanics. Bohm was an interesting character destined to have an interesting life. At that point he was an assistant professor of physics at Princeton University, specializing in plasma physics. He had come to Princeton from Berkeley, where he had been a student of J. Robert Oppenheimer. Like many people around Oppenheimer, he had been a communist sympathizer and briefly a Communist Party member before the war. As a result, the U.S. Army had refused Oppenheimer's request to bring Bohm along to work on the atomic bomb at Los Alamos.

There is no evidence that Bohm was ever a spy or a Soviet agent, but, like others with integrity, when called in 1950 to testify before the House Un-American Activities Committee, he asserted his Fifth Amendment rights and so avoided informing on others. He was arrested and charged with contempt of Congress, but acquitted. Princeton, to its shame, suspended, and then declined to renew, his faculty appointment.

Einstein proposed appointing him at the Institute for Advanced Study, but was unable to overcome opposition from its administration. At that very moment, when Bohm found himself unemployed and, in the United States, likely unemployable, his textbook was published to high praise.

There has been no shortage of textbooks published on quantum mechanics since the first, by Paul Dirac, one of the inventors of the theory, which appeared in 1930. Bohm's is one of the best. And despite persistent doubts over many years, when he discussed interpretational issues he kept close to the Copenhagen orthodoxy. One section of his book was titled "Proof that quantum theory is inconsistent with hidden variables." Another was about the "Unlikelihood of completely deterministic laws on a deeper level."

Einstein summoned him. He expressed his admiration for the

lucidity of Bohm's defense of the Copenhagen view, but asked for a chance to explain his point of view and perhaps change Bohm's mind.

It appears Einstein succeeded. After talking with Einstein, Bohm began to think about whether there might be a deeper theory, which was realist and deterministic. Perhaps it was the appeal of realism to a Marxist materialist; perhaps it was the clarity of Einstein's thinking. But it didn't take long for Bohm to invent a realist completion of quantum mechanics. What he did was, basically, to reinvent de Broglie's forgotten pilot wave theory.

There is, it should be mentioned, a difference between de Broglie's and Bohm's theories, in that Bohm chose a different law for the guidance equation by which the wave guides the particle. As I explained above, de Broglie's guidance equation has the particle taking the path of steepest ascent up the wave function. This determines the speed and direction of motion of the particle.

In Bohm's theory, the law that guides the particle is a version of Newton's law of motion: it describes how a particle accelerates in response to a force. What is new is that there is a force which guides the particle to move to where the wave function is largest. In addition, Bohm has to assume one more condition, which is that at the initial moment, the velocities of the particles are those given by de Broglie's guidance equation.

Apart from this difference, de Broglie's and Bohm's theories are different versions of the same idea, which is that both the wave function and the particles are real, with the waves guiding the particles. As presented originally, they are equivalent in that they predict the same trajectories for the particles. As a result, both theories predict that if an ensemble of particles starts off distributed according to Born's rule, that rule will continue to be satisfied as the wave function changes and the particles move around.

It didn't take long for Bohm to write two papers presenting his new theory.[1] He submitted them to the most prestigious journal at that time, *Physical Review*. He also sent drafts to several people, including de Broglie, who quickly published a short article explaining why Bohm's theory, like his own previous proposal, was wrong.

Bohm added a very interesting sentence to his manuscript: "After this article was completed, the author's attention was called to similar proposals for an alternative interpretation of the quantum theory made by de Broglie in 1926, but later given up by him."

This sentence certainly claims that he didn't know of de Broglie's pilot wave theory when he invented his own version. This in itself is a little shocking, given that de Broglie was a world-famous Nobel Prize winner, universally recognized for having proposed that electrons and other particles have waves. But there it is.

Bohm also devoted a section of his second paper to explaining why von Neumann's theorem doesn't apply to the theory he is proposing.

Bohm's first paper on the pilot wave theory appeared in January 1952. By then he had taken a professorship in São Paulo, Brazil. From that far remove, lonely and sick from the unfamiliar food, he waited as the responses to his revolutionary papers drifted in by letter.

One person Bohm might have hoped for support from was Einstein. The great savant had, after all, praised pilot wave theory when it was first presented by de Broglie. But, apparently, by the time Bohm published his papers, twenty-five years later, Einstein had changed his mind.

Einstein described his reaction in a letter to Max Born: "Have you noticed that Bohm believes (as de Broglie did, twenty-five years ago) that he is able to interpret the quantum theory in deterministic terms? That way seems too cheap to me."[2]

He went on, "This path seems to me too easy." It is a "physical fairy-tale for children, which has rather misled Bohm and de Broglie."[3]

Einstein elaborated in a paper in honor of Born, posing an objection. Bohm's theory predicts the motion of the particle, and one consequence is that in a stationary state of an atom, the electron is predicted to be simply standing still. As Einstein explained, "The vanishing of the velocity contradicts the well-founded requirement, that in the case of a macro-system the motion should agree approximately with the motion following from classical mechanics."[4] But it doesn't because, according to classical mechanics, the electron should be orbiting the nucleus, and not just standing still.

It should have been immediately apparent that Einstein's objection is wrong, because atoms are not "macro-systems." But nonetheless, Einstein's objection points to how different the particles of pilot wave theory are from those of Newtonian physics. As I stressed earlier, de Broglie had understood from the beginning that his particles would move in ways that violate basic principles of Newtonian physics, such as the principle of inertia and the conservation of momenta. This was necessary if light quanta could bend their trajectories to diffract around obstacles. De Broglie's and Bohm's guidance equations resulted in trajectories that diffracted and refracted, but there was a price to pay, which was apparent violations of basic principles. Particles that just stood still in an atom, and did not need to orbit to keep from falling into the nucleus, also contradicted these principles. For Einstein, it seemed, the price was too high.

Einstein's dislike of pilot wave theory didn't prevent him from writing sympathetically when he heard through a mutual friend about Bohm's "feeling of distress for being closed out and closed in at the same time. What impressed me most was the instability of your belly, a matter where I have myself extended experience."[5]

Indeed, the other responses Bohm received or heard about were not likely to have helped his digestion.

Heisenberg replied that from his operational point of view, the particle trajectories in Bohm's theory constituted an extraneous "ideological superstructure." There were two possible fates for any proposed alternative to quantum mechanics. Either the new theory gave predictions that disagreed with those of quantum mechanics, in which case it is most likely wrong, or it predicts the same phenomena, in which case it has nothing new to offer physics. He wrote that "Bohm's interpretation cannot be refuted by experiment. . . . From the fundamentally 'positivistic' (it would perhaps be better to say 'purely physical') standpoint, we are thus concerned not with counter-proposals to the Copenhagen interpretation, but with its exact repetition in a different language."[6]

Pauli issued a similar criticism, but after further study, conceded: "I do not see any longer the possibility of any logical contradiction as long as your results agree with those of the usual wave mechanics and as long as no means is given to measure the values of your hidden parameters."[7]

In fact, there are circumstances in which the predictions of pilot wave theory disagree with those of quantum mechanics, but it took some time for that to become clear. We will return to this point shortly.

Not everyone was so kind. Back in Princeton, Robert Oppenheimer declined to read Bohm's papers, calling them a waste of time. But this did not prevent him from pronouncing that Bohm's work was "juvenile deviationism."[8] Doesn't that sound exactly like language one Marxist would use to condemn another? Oppenheimer's last word on the subject was "If we cannot disprove Bohm, then we must agree to ignore him."[9]

The mathematician John Nash, now famous for his theorem on equilibrium in economics, wrote to Oppenheimer to complain about the dogmatic attitudes he found among the Princeton physicists, who treated anyone who "expresses any sort of questioning attitude or a belief in 'hidden parameters' . . . as a stupid or at best quite ignorant person." Nonetheless, he was with the other losers, because he confessed, "I want to find a different and more satisfying under-picture of a non-observable reality."[10]

The complete rejection of his breakthrough work by Oppenheimer, who had been both a mentor and a father figure to Bohm, must have hurt deeply. Bohm was doubly exiled from Princeton, then the center of American physics, for his rebellion against the Copenhagen philosophy and his simultaneous refusal to capitulate to the McCarthyist witch hunt. One must admire the courage that took, while remembering the cost. Bohm was isolated at what must have felt to him like the end of the Earth.

Bohm's friends and his biographer intimate that Oppenheimer had motives to distance himself from a suspected "red," as he was himself in danger, about to be caught up in the same witch hunt. But even putting that aside, it would be naive to believe that in the absence of his political catastrophe and exile, a Bohm who had stayed in Princeton would have succeeded any better in gaining interest in his subversion of the Copenhagen ideology.

In any case, the response from Copenhagen appeared equally dismissive. There is a report, by the philosopher Paul Feyerabend, who visited Copenhagen then, that Bohr was at least momentarily "stunned" by Bohm's papers. But if he was stunned it was not enough to ever mention in his own writings, let alone pick up a pen and respond to Bohm directly. Instead, Bohm received a letter from a protégé of Bohr named Léon Rosenfeld.

Here is a sample of Copenhagen-speak, taken from that letter:

> I certainly shall not enter into any controversy with you or anybody else on the subject of complementarity, for the simple reason that there is not the slightest controversial point about it. . . . [T]here is no truth in your suspicion that we may be talking ourselves into complementarity by a kind of magical incantation. I am inclined to retort that it is just among your Parisian admirers that I notice some disquieting signs of primitive mentality.
>
> The difficulty of access to complementarity which you mention is the result of the essentially metaphysical attitude which is inculcated to most people from their very childhood by the dominating influence of religion or idealistic philosophy on education. The remedy for this situation is surely not to avoid the issue but to shed off this metaphysics and learn to look at things dialectically.[11]

Reading this alone in his São Paulo apartment, David Bohm must have felt a long way from Kansas, or, in his case, Pennsylvania.

Despite his disappointments, Bohm was productive during his time in Brazil. He continued to make contributions to plasma physics while he focused on his new quantum theory, and he began a collaboration with Jean-Pierre Vigier, a student and colleague of de Broglie. But he was not happy in Brazil and in 1955 moved to the Technion in Israel, then a few years later to England. After a stay in Bristol he ended his odyssey at Birkbeck College, University of London, where he was to stay for the rest of his life.

In London, Bohm moderated his communist sympathies; like many who had given the Soviet Union the benefit of the doubt, he was shocked as the thawing of Soviet power under Nikita

Khrushchev led to the confirmation that the Stalinist gulag had indeed been every bit as murderous as had been reported. Bohm's desire for a road to the perfectibility of human beings then turned to mysticism, and after a short immersion in the teachings of the mystic Gurdjieff, he fell under the influence of the Indian guru Krishnamurti.

Bohm meanwhile continued his relentless search for a deeper viewpoint on nature that would take him beyond the quantum theory. This led him to develop a highly original line of thought, frankly speculative and philosophical, both related to and transcending his physics. He wrote several books, through which he reached a new audience of artists, philosophers, and seekers, while his dialogues with Krishnamurti became very popular in the wider world.

ALTHOUGH HIS LATER WORK is of no relevance for judging the importance of his work on pilot wave theory, I do feel it would be irresponsible and cowardly to not attempt a summing up of the life's work of this complex and contradictory sage. I feel a genuine sympathy for David Bohm in his search for transcendence, first through the Marxist vision of a new human psychology arising from the dream of a just and equal society, and then, when that fantasy was exposed as a cruel illusion, through his work with mystics.* From

* If I can be permitted a purely personal remark, I am a grandchild of a Marxist who remained a lifelong member of the American Communist Party long after the dream had died, and I am also the son of seekers who spent many years in the Gurdjieff work. To a large extent, the errors of my parents and grandparents inoculated me against falling in love with organized seekers, running after visions of transcendence. It is easy for me to criticize Bohm and others of his generation for the astounding naiveté they showed in the face of the peculiar combination of genuine compassion for human suffering and extraordinary dishonesty and narcissism that gurus like Gurdjieff and Krishnamurti shared with the "revolutionary" leaders on the vanguard of the left. But at the same time, there is, I believe, the shadow of something real behind the teachings of the likes of Gurdjieff and Krishnamurti, who brought distillations of Eastern spiritual practices to westerners.

Oppenheimer to Krishnamurti, some weakness in Bohm made him susceptible to that kind of dominating, supremely confident figure.

But as much as one can criticize Bohm for what in retrospect looks like the naive and ignorant suspension of his better judgment, his years of hard, determined effort in search of the science beyond the quantum rescues his life's work and restores to it integrity, seriousness, and promise. He was on a quest for a new transcendent form of science, informed simultaneously by the deepest strands of what is best called religious thought and the knottiest puzzles of theoretical physics. It is a domain few good physicists have explored; perhaps only David Finkelstein can be mentioned here. It is easy to say that Bohm failed, and that his greatest achievements by far were his early contributions to quantum physics. At the same time, he explored a road that few of us have had the courage or the vision to even take one step toward, in spite of the obvious fact that the greatest dangers we face as a species can be tied to the utter incoherence of human culture, a break that has its roots in the incommensurability of scientific and spiritual understandings of the world.

IN THE WAKE OF what we've learned from Bohm, let's sum up. The pilot wave theory explains everything that ordinary quantum mechanics does, without the awkwardness of Rule 2. The wave function evolves always according to Rule 1, so it never jumps or collapses. What is new is that there is a particle that moves according to its own law, guided by the wave function. Together the two laws give an entirely realist description of quantum phenomena.

In addition, pilot wave theory explains what quantum theory does not. It gives a complete description of what goes on in every individual process. It explains how and why electrons move. It explains where the uncertainties and probabilities come from, which is our ignorance about the starting positions of the particles. And it solves the measurement problem because there is no need to distinguish experiments from other processes.

In the second paper Bohm wrote in 1952 on the new theory, he studied the measurement process in detail and showed that, in the case of an atom interacting with a detector set up to measure some property of it, the detector ends up correlated with the atom, in terms of where the particles are as well as in terms of the wave functions. Thus, measurements work correctly on both sides of the double ontology.

From a realist point of view, pilot wave theory is vastly superior to the Copenhagen interpretation. By its very existence it gives the lie to Bohr and Heisenberg's argument that it is impossible to have a realist description of quantum physics. One might have thought that the community of physicists would have jumped to embrace pilot wave theory, either when de Broglie first proposed it to the Solvay conference in 1927 or in 1952 when Bohm proposed it again. This is clearly what Bohm expected, and his disappointment, as he waited in São Paulo, may be ours as well.

Some historians have suggested that the embrace of anti-realism by the European physics community in the 1920s was part of a larger cultural movement which embraced irrationality as a response to the slaughter in the trenches that their generation had recently experienced. But this does not explain the similar rejection of pilot wave theory by the physics community of the 1950s, which had recently come to be dominated by the triumphant, optimistic, and pragmatic American spirit.

Some might explain it by the power of research schools led by charismatic leaders, particularly Niels Bohr, who inspired and mentored many of the quantum revolutionaries who came from across

Europe and America to work with him. De Broglie, by contrast, had just a few students throughout his long life, and they were, to my knowledge, all French. His small group of acolytes was isolated even within the community of French physicists.

Bohm inspired the development of a community of theorists in Brazil, for which he is unappreciated beyond that country. After Brazil, in Israel and London, he had a few good students, one of whom, Yakir Aharonov, became a leading theorist with his own ideas and program, quite different from Bohm's. A handful of Bohm's students became specialists in quantum foundations, but they pursued diverse ideas and did not form into a coherent Bohmian school of thought. It didn't help that, by the time Bohm had relocated to London and was back in a place where he could assert influence, much of his attention was captured by mysticism.

Nonetheless, interest in pilot wave theory grew slowly but steadily over the years, as it was taken up and developed by a small number of good scientists around the world. By the 1990s, what was sometimes called "Bohmian mechanics" constituted a small but distinct and visible subculture of the community of scientists, mathematicians, and philosophers who devoted themselves to the puzzles of quantum foundations.

DUE TO THE WORK of these "Bohmians," some subtle questions about pilot wave theory have been raised and answered. One of the trickiest questions has to do with how probabilities arise in pilot wave theory. The theory is deterministic. Given a wave function at one time, we can use Rule 1 to determine the wave function at any future time. The equation that describes how the wave function guides the particle is also deterministic, and if we specify where the particle is at an initial time, it will tell us exactly how the particle moves from then on. Each particle has a definite trajectory.

So where do probabilities come from? Probabilities enter for the same reason they can enter Newtonian physics: because of our ignorance about the exact positions of the particles. As we cannot know where the particle starts out, we are uncertain about where it will be in the future. Probabilities in pilot wave theory express our ignorance of where the particles were initially.

To make sense of probabilities in pilot wave theory, we have to picture a collection of systems with the same wave function, but with different starting positions of the particles. The particles are distributed initially according to a probability distribution function, which tells us how common the different initial positions are in the collection.

We are free to choose the initial positions of the particles, to make the probability distribution function be anything we like. We evolve the system forward in time, using Rule 1 to evolve the wave function and the guidance law to move the particles around. When we do this, the probability distribution function changes in time as well, reflecting the particles moving around.

In quantum mechanics, as I described earlier, the probabilities of finding the particles in different places are given by Born's rule to be the square of the wave function. That is simply posited in quantum mechanics as part of Rule 2. In pilot wave theory the particles have their own reality, and we are free to choose the initial probability distribution function. One choice we can make is that it is given, just as in quantum mechanics, by Born's rule. To do this, we distribute the particles so that the larger the square of the wave function is, the more particles in the collection are placed there.

When we make this choice, it is maintained in time. The particles move around and the wave function changes in time, but it remains true that the square of the wave function gives the probability of finding a particle.

But in de Broglie's formulation, there is more. Suppose one starts the collection off with a different probability distribution for the particles, one not given by the square of the wave function. Then the system will evolve in a way that brings the actual probability distribution into agreement with that given by the square of the wave function. This was shown in an important result of Antony Valentini's.[12] It has been confirmed by numerical simulations since.[13]

This is analogous to how thermodynamics works. When a system of many particles is in equilibrium with its surroundings, the entropy is maximal. This is because entropy is a measure of disorder, which typically increases over time. If one starts the system off in a different configuration, one more ordered than equilibrium, it is most probable that the disorder will increase until the system is in equilibrium.

The case of de Broglie's pilot wave theory is very similar. We can say that a quantum system is out of *quantum equilibrium* if the probability distribution for where a particle might be found is different from that given by the square of its wave function. When they agree, the system is in quantum equilibrium. Valentini's theorem tells us that a quantum system out of quantum equilibrium is most likely to evolve until it reaches the state of quantum equilibrium.

Once a system is in equilibrium, the predictions of pilot wave theory agree with those of conventional quantum mechanics. Thus, one has to somehow drive a system out of quantum equilibrium to set up a situation in which an experiment could distinguish pilot wave theory from quantum mechanics.

Physics out of quantum equilibrium contains several surprises.

One is that it becomes possible to send information faster than light. This is a consequence of another result of Valentini's, which tells us that while the system is out of quantum equilibrium, information and energy can be sent instantaneously, contradicting special relativity.[14] Needless to say, if this were to be confirmed experimentally it would be of the first importance for our understanding of nature and possibly even for technologies that science fiction writers dream of. This is one way an experiment could very dramatically distinguish pilot wave theory from conventional quantum mechanics. There have been a few attempts to drive quantum systems out of quantum equilibrium and test these predictions, but, so far, they haven't succeeded in either discovering quantum nonequilibrium or ruling out pilot wave theory.

One place to look for out-of-quantum-equilibrium physics is in the very early universe. Valentini and collaborators have hypothesized that the universe began in the big bang out of equilibrium, and equilibrated as it expanded. This might have left traces in the cosmic microwave background, or CMB, which are being searched for, but there is no clear evidence yet.[15]

LET'S COME BACK TO the Schrödinger's cat experiment and see how pilot wave theory resolves it.

Pilot wave theory asserts that quantum mechanics applies universally. There is only Rule 1, and it applies to all cases. This means that measurements are no different from other processes.

Everything—atoms, photons, Geiger counters, cats, and people—has a dual existence, as a wave and a particle. Both sides of this double existence are complex for large complicated objects such as Geiger counters or cats, which are made of many particles working together. We need a word to talk about all the ways that the particles making up a cat may be arranged in space, and we have one: the *configuration* of the atoms. If you speak of where all the atoms making up the cat are located with respect to each other, you are describing the configuration of the cat. Because there are many atoms, it takes a great deal of information to describe the cat's configuration.

All this information must be coded into a list of numbers. How many numbers does it take to describe a cat? For just one atom it takes three numbers. These locate the atom in three-dimensional space. For two atoms it takes six numbers, three for each atom. So to locate the atoms in a cat takes three numbers per atom. There are very roughly 10^{25} atoms in a cat, so it takes 10^{25} multiplied by three to describe the cat's configuration.

The important thing about pilot wave theory is that the atoms are all real, and they are each located somewhere definite in space. Each atom has a location, which is a point in space. Each cat has a configuration, which amounts to saying that each of its atoms is located somewhere definite in space.

An atom also has a wave, which is located in three-dimensional space. Each cat also has a wave associated with it. The strange thing is where that wave is located. It isn't a wave in three-dimensional space. Instead, it's a wave in a very high-dimensional space, called the configuration space (see figure 9). Each point of this space corresponds to a configuration of the cat.

It is difficult, if not impossible, to visualize a space with many dimensions. I once watched in awe as Roger Penrose did a calculation on the blackboard which required him to slide a two-dimensional surface around a six-dimensional obstacle in an eight-dimensional space, and I did have the thrilling experience of following step by step, but that's the limit of my experience. Most mathematicians are not as visually gifted, but we can reason our

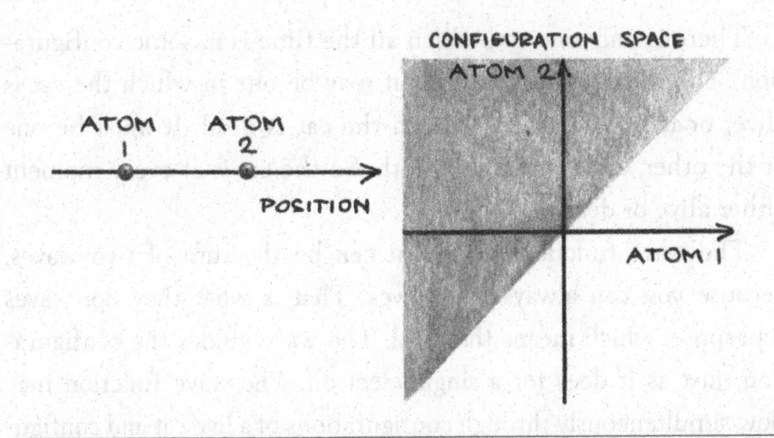

FIGURE 9. CONFIGURATION SPACE Two atoms live on a line, in one dimension. Their configuration is measured by two numbers, so their combined configuration space is a point on a plane, in two dimensions. We treat the two atoms as identical, so atom 2 is always the rightmost atom.

way around in a high-dimensional space. When I draw a three-dimensional object, I am really drawing a two-dimensional projection of it. Likewise, what I see in my mind when I imagine a configuration space like that of a cat, with perhaps 3×10^{25} dimensions, is a three-dimensional projection, together with a silent admonishment to be careful and not draw false conclusions from this totally inadequate visualization.

A wave on the configuration space carries a vast amount of information. Recall, for example, the state CONTRARY, which describes correlations between the answers to questions asked simultaneously of two particles, while telling nothing at all about each particle separately. To code quantum states like this, in total generality, we need more than a three-dimensional wave for each atom in the cat. We need a wave flowing on the space of all possible configurations of the cat.

Once one accepts the existence of a wave on the space of all the configurations of a cat, the resolutions of the quantum puzzles follow directly.

There is only one cat, which all the time is in some configuration. The configuration of the cat may be one in which the cat is alive, or it may be one in which the cat is dead. It must be one or the other, but it cannot be both. So the cat is at every moment either alive or dead.

The wave function of the cat can be the sum of two waves, because you can always add waves. That is what they do: waves superpose, which means they add. The wave guides the configuration, just as it does for a single electron. The wave function may flow simultaneously through configurations of a live cat and configurations of a dead one. Just as a river can branch, and take both branches, a wave function may branch and take both the branch over living configurations and the branch over dead configurations.

The wave function ends up related to the probability of finding different configurations. When the wave function is large over some configuration, so is the probability. So the probability of finding the cat in a live configuration or a dead configuration may be each roughly one half. But there is only one cat, and just as an electron can be in only one place at a time, the one cat is either dead or alive.

Is it weird that the wave function will spawn branches that flow to where the particles or their configurations are not? A bit, but this must be, because the particle can follow only one branch. But an empty branch may have consequences in the future. The different branches may flow back together in the future, making interference patterns that influence where the particles go.

Eighteen years ago I had a difficult decision to make. Two futures beckoned to me, each of which seemed, from all the information I could gather, attractive. Of course one never has enough information to make a decision like this, for everything about my future was at stake. The question was, Which country and which city would I call home? Whom I might marry, who my children would be, what languages they might speak, and how long I might live would all be influenced by this decision.

Unable to decide, I consulted friends with a quantum lab and let a radioactive atom decide for me. If the atom decayed within its half-life, I would take the new opportunity in a new city and country; if it didn't, I would stay with the familiar. Well, it decayed, and here I am in Toronto, with my family, friends, and neighbors, none of whom I would ever have met had that atom held off decaying a bit longer.

There is nothing special about me. All of us are made of particles that have been guided to the present by a wave function on our vast space of possible configurations.

The wave function surrounds where I am now, but it also has other branches where I might be, but am not. Some of them branch from that experiment and develop the empty history that I did not take, but would have had the atom not decayed. The empty wave function of the particle flows on from there, to this day. This empty branch of my wave function continues to inhabit London.

Have we not all felt a bit wistful contemplating lives we might have had, had a decision turned out just a bit differently? If the pilot wave theory is right, then these lives not taken are traced by an empty wave function, ready to guide my atoms, which, however, are elsewhere.

A few years before that, my wave function faced another fork, from which two very different branches flowed. I took one branch, but had I taken another, I would have faced a very different fate.

I was booked on a Swissair flight from New York to a conference in Vienna. The night before my trip, I heard from the organizers that my talk wasn't scheduled till the end of the meeting, and so, on a total whim, for a reason I no longer remember, I called the travel agent and rebooked for a day later. Before going to sleep the next evening I turned on the radio and heard that the flight I would have been on had crashed off Halifax. So, if pilot wave theory is correct, it is really true that a branch of the wave function of the atoms that then constituted me is to this day bunched up at the bottom of St. Margarets Bay, off the village of Peggy's Cove, Nova Scotia.

That branch is empty, as are myriad others. But if pilot wave theory is right, they are real. The only difference between them and the one branch that guides me now is that only one branch coincides with, and guides, the atoms that make me up. The myriad other branches flow on, empty.

Do I care about these other branches? Should I? There is always the chance that at some time in the future an empty branch recombines with my branch, causing interference, which changes my life abruptly.

The chances for this to happen are extraordinarily small. They are in a category of possible events that would be permissible under the laws of physics but which essentially never happen. All the atoms in the air in the room where I'm typing this might by chance line up together and fly out the window, asphyxiating me. But this would be extraordinarily unlikely, given that the atoms spend their day bouncing around randomly.

So there is basically no chance that the empty branches representing the lives we didn't live and the choices we didn't make will have any effect on our futures. But were we mere atoms, interference between full and empty branches of the wave function would be happening all the time.

So for all practical and moral purposes, if pilot wave theory is right, we can ignore the empty branches. We are real only once, and live out that life on that one occupied branch. We need care about, and be responsible for, only what the one real version of each of us does.

NINE

Physical Collapse of the Quantum State

Experiment and common sense suggest that there are no superpositions of macroscopic objects, because every large body is always somewhere particular. Rule 2 was invented to accommodate this, at least as regards the behavior of measurement instruments and the systems that come into contact with them. To avoid superpositions of the states of a measuring instrument, Rule 2 dictates that just after a measurement of a particle's position, its wave function immediately collapses to a state corresponding to the position that was measured.

Just before the measurement, a certain atom's wave function might have been spread all around the Earth, giving it an equal probability of being found anywhere on the globe. But when a measurement is done of its position, and if that measurement reports the atom's location to be somewhere in New York City, then, at the moment that report is made, the atom's wave function collapses down to the extent of the five boroughs.

In standard quantum mechanics this collapse of the wave

function happens only as the result of a measurement. This raises a problem for realism, because it is only our use and interpretation of the result that determines whether an interaction with a large body is a measurement or not.

According to a realist perspective, a measuring instrument is a physical system, which happens to be large, and which has a special capacity to amplify tiny differences in an atom's behavior to make a record of what was seen that can be etched in a macroscopic change. But because it is a physical system it should obey the same laws as the atoms which compose it. If the atoms can be in superpositions, the same should be true of the vast collections of atoms that make up the measuring instrument. In the last chapter, we saw that in pilot wave theory, part of the price we pay for realism is a world full of empty branches of wave functions, which have long since disconnected from the objects they might guide.

But what if the collapse were a real physical process that occurs whenever a large body is involved in an interaction? The collapse would be triggered by the size of the object, measured in mass or in the number of atoms that make it up, irrespective of its use as a measuring instrument. The wave functions of all large bodies would collapse, wiping out their superpositions. The measurement systems, made of myriad atoms, would collapse too. This suggests a strategy for a realist version of quantum physics.

The idea would be to modify quantum mechanics by combining Rule 1 and Rule 2 into a single rule, which specifies how wave functions evolve in time. When the system it is applied to is microscopic, the old Rule 1 is a good approximation. Collapses of the wave functions of atoms may happen, but only rarely. But when the system is large, collapse happens frequently, so that it appears that the body is always somewhere definite.

Theories of this kind have been constructed since the 1960s; they are called *physical collapse models*.

The first physical collapse model was invented in 1966 by Jeffrey Bub, a student of David Bohm, and developed by the two of them.[1] In the same year, F. Károlyházy published a paper arguing that noisy fluctuations in the geometry of spacetime could cause the wave function to collapse. As with pilot wave theory and Bell's work of the same period, the response to these pioneering papers was slow. The first person to develop a completely precise version of a theory of this kind was Philip Pearle, an American theorist who has done very important work in spite of spending his career at a small undergraduate college. He struggled for almost a decade to invent a consistent theory for physical wave-function collapse, and his first theory incorporating a physical collapse of the wave function was published in 1976.[2]

Pearle's version of a collapse model adds a random element, so that there is something akin to a roll of the dice that decides when and where a wave function collapses. The rolls are infrequent for wave functions of atoms, and so small systems consisting of a few atoms collapse infrequently. But collapse occurs often for macroscopic systems containing many atoms. Pearle called his theory *continuous spontaneous localization*, or CSL.

For several years Pearle was nearly the only one working on this approach to realism. Then in 1986 three Italians working in Trieste proposed a rather elegant version of the idea, which has been known since as the GRW theory after their names, Ghirardi, Rimini, and Weber.[3] Other people joined in to develop these dynamical collapse models, including Lajos Diósi, Lane Hughston, and Nicolas Gisin.

These theories differ from each other at the level of details, but they share the key feature that the behavior of any quantum system is a mixture of Rule 1 and Rule 2. Most of the time the wave function of an atomic system changes slowly and smoothly, following Rule 1. But from time to time it jumps abruptly into a definite state, following a form of Rule 2.

One defect of these spontaneous collapse models is that the rate of the spontaneous collapses has to be carefully specified so that the collapses are rare enough not to corrupt interference patterns built by delicate superpositions in atomic systems. This guarantees the successes of quantum mechanics by preserving the coherence of superpositions of microscopic systems, where it is needed. But the wave function of a large body will get hit with a collapse far more often, because it consists of many atoms. Events that are rare for one atom will happen frequently to some atom or another in a large collection of them. But when one atom collapses, so must the others making up the same body. As a result, the model can be tuned so that the wave functions describing macroscopic systems collapse far more frequently, explaining why large-scale objects are always somewhere. This solves the measurement problem.

These theories have no need of particles, in the sense of pilot wave theory. There are only waves, but the result of a spontaneous collapse will be a wave highly concentrated around one location. Such a concentrated wave is hard to distinguish from a particle.

Because there are no particles, the mysteries of the wave-particle duality evaporate. One just has to understand why waves evolve under two very different processes.

These collapse theories are entirely realist. The wave function *is* the system, and there are no mysteries as to how to interpret it. By collapsing the wave function down to only what is physically relevant, collapse theory avoids the extravagant proliferation of branches that burden the pilot wave theory. There is no measurement problem because big objects, including measuring devices, are always in collapsed states. There is no special role for consciousness, information, or measurement. What you see is what you get.

To define one of these theories you have to decide which of the incompatible questions the collapsed wave function is to answer. The usual answer is position in space. The collapsed wave functions are peaked somewhere in space, which makes them like particles.

One consequence is that energy is no longer precisely conserved. A metal block should slowly heat up as a result of all the collapses the wave functions of its atoms undergo. This, for me, is the least attractive feature of spontaneous collapse models. On the plus side, there are experiments planned to look for this heating.

As is often the case with new theories, there is a lot of freedom. One is free to adjust how often the collapses take place. One can make this rate depend on the mass or the energy of the atoms. If the hypothesis of spontaneous collapse is to be viable, there must be a way to set the rates so that wave functions of atoms and elementary particles rarely collapse, while big things collapse often enough that they are always in some definite place. And one has to make sure all unintended consequences, such as heating up of matter, are undetectable. Remarkably, these conditions can all be met, so these theories are viable.

In some of these models the spontaneous collapses are random processes. The theory specifies only a probability for collapse to happen. This leads to uncertainties and probabilities, which are built in from the beginning. The probabilities are coded into the fundamental laws, rather than being a consequence of ignorance or belief. The intrinsic randomness of the collapse process then explains the uncertainties in quantum physics, and it does so in a way that does not single out measurements. Thus, the probabilities are explained in a way that is perfectly compatible with realism. That is a great advantage. (Of course, if one wants a deterministic theory, this is a disadvantage.) Related to this is the fact that the fundamental laws are irreversible, so that the arrow or direction of time is coded in at the bottom level. Some may see these as defects, but my view is that they are very positive features of the collapse models.

One worrying aspect of spontaneous collapse models is that the collapse of the wave function takes place all in one moment of time. As the wave function may be spread out over space, its collapse defines a moment of simultaneity over a whole region. This appears to contradict relativity theory, which asserts that there is no physically meaningful notion of simultaneity over regions of space. While this does seem to be a problem for the original dynamical collapse models, there have been proposals for making collapse models that are consistent with special relativity.[4]

But the most attractive feature of all the collapse models is that they predict new phenomena, which are subject to experimental testing. The random collapses introduce noise into a system. For some values of the parameters, the effect would be large enough to be seen. No need for such a noise source has been seen in several recent experiments, which rules out certain values of the parameters, if not the theory itself. This is real science, and the experiments continue. Nothing would be more wonderful than the discovery of an effect which contradicts quantum mechanics and confirms a prediction of one of its realist alternatives.

One weakness of some of these collapse models is that they make no reference to, or use of, other key questions of physics. It would be more compelling if the modifications we make to quantum mechanics are motivated by a problem besides the measurement problem, such as the problem of quantum gravity. This brings us to the work of Roger Penrose.

IF THERE IS ONE living theorist whose achievement and depth of insight and influence match those of the sages of the early twentieth century, it is Roger Penrose. Simply put, he is the real thing.

Penrose follows his own compass and he has, as a result, novel and surprising things to say about most issues in fundamental physics, including quantum gravity and quantum foundations. Because everything he envisions is tied together by an often-hidden consistency, the best way to approach his proposal for quantum theory is by tracing his work back to his time as a young mathematician in the early 1960s, when he was fascinated by the foundations of our understanding of space, time, and the quantum.

It is easy, but inadequate, to describe Penrose as the most important contributor to general relativity since Einstein. In the early 1960s he invented revolutionary new mathematical tools to describe the geometry of spacetime, based on causality. Rather than talking about how far away two events are, or how much time elapses on a clock, he described spacetime in terms of which events were the causes of which events. This led him to posit and prove theorems that showed that, if general relativity is right, the gravitational field becomes infinitely strong within the core of black holes.[5] Once that happens, the theory breaks down, because its equations stop working to predict the future. Such places, where time may start or stop, are called singularities. Afterward, working with Stephen Hawking, he extended his method to the expanding universe and proved that general relativity predicts that time began a finite time into our past, when the whole universe began its expansion in a state of infinite density.[6]

But his inventions exceed even these transformative contributions to general relativity. Like Einstein, Penrose cares more deeply than most for the coherence of our understanding of the world. And, just as it did in the cases of David Bohm and David Finkelstein, this passion has driven Penrose to develop a unique vision of fundamental physics, which is unmistakably his. Moreover, Penrose's vision has, over the many years of his creative career, led him to invent mathematical structures that others later utilized.

After transforming the practice of general relativity, Penrose turned his attention to fundamental physics. He was struck by a sympathy between quantum entanglement and Mach's principle—the idea, which had inspired Einstein's invention of general relativity, that what is real in general relativity is relationships. Both ideas hint at a global harmony which ties the world together.

Penrose was the first to ask whether the relations which define space and time could emerge from quantum entanglement. Seeking insight into this question, he was inspired to invent a simple game based on drawing diagrams, the rules of which represented simultaneously quantum entanglement and aspects of physical geometry. This game, his first vision of a finite and discrete quantum geometry, Penrose called *spin networks*.

Most theoretical physicists work out their ideas by doing calculations in existing theories. Penrose works sometimes instead by inventing games. Their simplicity captures profound questions, which one investigates by playing the game. It is typical of Penrose that his main paper on spin networks was not only unpublished—it was never even typed up. Mimeographs (now they would be called photocopies) of his handwritten notes circulated among his students and from friend to friend. These notes were an exhilarating read, even though they ended in the middle of the main proof.*

For decades spin networks remained a kind of philosophical

* Which was completed in a PhD thesis of a student of Penrose's called John Moussouris, which also remained unpublished, and was also passed hand to hand.

parlor trick, passed on by sketches on napkins during dessert at conference dinners. But they turned out, years later, to be the central structure in an approach to quantum gravity called *loop quantum gravity.* In that context, spin networks embody one way that the principles of quantum theory and general relativity can coexist.

Extending spin networks, Penrose discovered *twistor* theory, which is an extraordinarily elegant formulation of the geometry underlying the propagation of electrons, photons, and neutrinos. Intrinsic to twistors is a beautiful asymmetry of neutrino physics, which is called *parity.* We say that a system is parity symmetric if its mirror image exists in nature. We have two hands, which are each mirror images of the other, so our hands are parity symmetric. But overall humans are not parity symmetric, because our hearts and other internal organs are arranged asymmetrically, and we each tend to favor one hand. Neutrinos exist in states whose mirror images don't exist, and hence are parity asymmetric. Penrose's twistor theory expresses this feature of neutrinos, because it uses mathematical structures which are not the same when looked at in a mirror.

For many years Penrose and a few students developed twistor theory, working in isolation in Oxford. In the late 1970s this caught the attention of Edward Witten, who many years later made twistors the keystone of a powerful reformulation of quantum field theory he invented with some younger theorists, which is still in progress.

What I find so remarkable about Penrose is that he has an inner narrative that connects everything he does into a coherent story. So it's no surprise that his expansive vision of a new physics led him to a reinvention of quantum mechanics. This was part of a larger strategy to combine quantum theory with general relativity, to make a quantum theory of gravity.

Typically, Penrose started off his attack on quantum gravity by ignoring the obvious path taken by nearly everyone else. The standard path is to construct a quantum description of a system, a process called *quantization*. This starts with a description of the system given in the language of Newtonian physics. We "quantize" this by applying a certain algorithm. The details of this don't concern us here, but suffice it to say the output is a quantum theory which is absolutely conventional and standard.

This technique works in many cases to give us successful quantum theories of atoms, elementary particles, and radiation. It can be applied to gravity; indeed, loop quantum gravity was made by "quantizing" general relativity.

Penrose took a different road. Quantum theory and general relativity clash on a few key points. The most crucial is that they have deeply different descriptions of time. Quantum mechanics has a single universal time. General relativity has many times—if by time we mean duration as measured by clocks. The beginning of Einstein's theories of relativity is a discussion of synchronizing two clocks. You start off by synchronizing them, but they do not generally stay synchronized. They slip out of synchronicity at a rate that depends on their relative motions and relative positions in the gravitational field.

Another point on which the two theories clash is the superposition principle. As we discussed, given two states of a quantum system, we can make new states by adding them together. Something we haven't needed to mention so far is that we can make a lot of different states from superposing the same two states. We do this by varying the contribution of each state to the superposition. Thus we can superpose CAT and DOG (from our earlier example) equally, as in

STATE = CAT + DOG

or we can choose instead

STATE = 3 CAT + DOG

or

STATE = CAT + 3 DOG

The number we multiply each state by is called an *amplitude.* Its square is related to the probability. Hence in the state CAT + DOG you are equally likely to find a cat lover or a dog lover, while someone in the state 3 CAT + DOG is nine times more likely to love cats than dogs.

General relativity does not have a superposition principle. You cannot add two solutions to the equations of the theory and get a new solution. Math-speak for this is to say that quantum mechanics is linear, while general relativity is nonlinear.

These two differences are related. The superposition principle is possible in quantum mechanics because there is a single universal time that we can use to clock how its states evolve in time. On the other hand, because distant clocks go out of sync, there is no simple way to add or combine two spacetimes to make a new spacetime.

Penrose embraces the multi-fingered nature of time in general relativity and the absence of superpositions as home truths. He suspects that the superposition principle must be violated once quantum phenomena are described in the language of general relativity. The simplicity and linearity of the superposition principle, he suspects, are only approximately true, and hold only to the extent to which the role of gravity can be ignored.

Thus, Penrose objects to quantizing gravity. Instead he suggests we should try to "relativize the quantum." By this he means to introduce the multi-fingered notion of time into quantum theory by violating the superposition principle and making quantum states nonlinear.

Penrose is a realist, but he makes an unusual move for a realist on quantum theory. Rather than ascribing reality to both waves and particles, or inventing new "hidden variables," Penrose takes reality to consist of the wave function alone. This leads him to take up the suggestion by Pearle and others that the collapse of the wave function during a measurement is a real physical process. The sudden change of the wave function is not, as some hold, due to an update in our knowledge of where the particle is; it is a genuine physical process.

Penrose, following the earlier work of Pearle and of GRW, proposed that collapse of the wave function is a physical process that occurs from time to time,[7] interrupting the smooth changes mandated by Rule 1. And he took up a suggestion made by Diósi and Károlyházy: that the collapse process has something to do with gravity.[8] When a wave function collapses, superpositions are wiped out. The rate at which a system's wave function collapses depends on the size and mass of the system. As we discussed earlier, this rate can be specified so that atomic systems almost never collapse, while macroscopic systems collapse often, so that superpositions of large objects are impossible.

What is really exciting about the work of Diósi, Károlyházy, and Penrose is that they proposed a criterion for when collapses would take place that makes the collapse an effect of gravity.

Roughly speaking, a superposition of an atom being here or there is collapsed to one location when the location of the atom would become measurable by the effect of its gravitational attraction.

This relates to the many-fingered time of general relativity. Imagine that the wave function is a superposition of an atom being in the living room with the atom being in the kitchen. Wherever the particle is, its mass has a gravitational field which affects clocks. One of the most striking predictions of general relativity is that clocks deeper in a gravitational field appear to slow down. This is well tested. Atoms on the surface of the sun have been observed to vibrate more slowly than the same atoms do on Earth. The effect is even seen by comparing the rates at which atomic clocks in the basement of a building tick compared to clocks on the roof.

The result is that clocks in the room where the particle is run slower than clocks in the other room. But what of a state which is a superposition of the atom being in the living room and in the kitchen? This seems to imply that the gravitational field must be in a superposition of states such that each clock runs slow.

But there is no such state, because one cannot add spacetime geometries to get new spacetime geometries. Hence the wave function must collapse.

Penrose gives a prediction for when the wave function will collapse, and work is underway to build an experiment to test Penrose's prediction. Very recently two experimental teams[9] have proposed that they may be able to construct superpositions of different gravitational fields, contrary to Penrose's hypothesis. This is fabulous, but what is worrying is that Penrose has not put forward a detailed theory unifying gravity and quantum theory from which his heuristic model can be derived.

Penrose has at least proposed a model for how this might all work, which combines the usual evolution of the quantum state, given by Rule 1, with collapse of the wave function, given by Rule 2. They go together into a single evolution rule.

Penrose's theory is not quantum mechanics; it is a new theory, which contains quantum mechanics within a realistic framework, based on a new evolution law, called the Schrödinger-Newton law. This unifies Rule 1 and Rule 2 into a single dynamical law.

If we focus on the behavior of atoms and radiation, this single evolution law mimics standard quantum mechanics. The superposition principle is satisfied to a good approximation. The wave function behaves like a wave, and Rule 1 is satisfied. Schrödinger's equation for the wave function is then recovered for atomic systems.

But if we pull back to describe the macroscopic world, Penrose's model describes a wave function which is collapsed and concentrated on single configurations. These concentrated wave functions behave like particles. So on the macroscopic level, Newton's laws for the motion of particles are recovered.

Thus, in the microscopic regime, this theory reproduces quantum mechanics, while in the opposite situation, it predicts that macroscopic objects behave like particles and obey Newton's laws.

Physical collapse models continue to be developed. Recently Pearle has made progress constructing a collapse model consistent with special relativity.[10] The idea that gravity is responsible for causing the quantum state to lose coherence, and hence collapse, has also been developed by Rodolfo Gambini and Jorge Pullin, who call their proposal the Montevideo interpretation of quantum mechanics.[11] And Steve Adler has found a role for spontaneous collapse in a hidden variables model he has been developing.[12]

Pilot wave theory and the collapse models have given us options for quantum physicists who want to be realists. The differences are striking, but so are the similarities.

One option is to believe there are both waves and particles; this leads to pilot wave theory. This easily resolves the measurement problem, but at a cost. The pilot wave theory is doubly extravagant. It has a doubled ontology, but an asymmetric dynamic by which the wave function guides the particles without there being any reciprocal action by which the particles influence the wave. And we have to live with a vast world in which the wave function has many empty, ghostlike branches.

The collapse models avoid all these objections. There are only waves, so there is no doubled ontology and no issue with reciprocation, and there are no empty branches because they are eliminated by the collapses. This also solves the measurement problem, but here, too, there is a price, which is that the theory comes with new adjustable parameters that must be tuned to keep the theory out of harm's way.

Both approaches agree on two key lessons: the wave function is an aspect of reality, and there is a tension with relativity theory. These are vital clues for the future of physics.

TEN

Magical Realism

Every quantum transition taking place on every star, in every galaxy, in every remote corner of the universe is splitting our local world on earth into myriads of copies of itself.

—BRYCE DEWITT

We saw in the last few chapters that there are options for realists, but notice that they all require changing the theory. The spontaneous collapse models make the sudden collapse of the wave function part of the dynamics of the theory. The collapse occurs whether or not measurements take place, and without regard to what we know. The resulting theories disagree with quantum mechanics generally, but preserve a domain of agreement sufficient for them not to contradict the results of experiments done so far.

Pilot wave theory is another option for realists. Rule 2 is suspended, so the wave function evolves always according to Rule 1. But a new element is added: particles, whose travel is guided by the wave function. So this theory is also different from quantum mechanics. When the particles are in quantum equilibrium, the predictions of the two theories overlap, but out of quantum equilibrium, the predictions of pilot wave theory differ from those of quantum mechanics.

It would be wonderful if someday experiments confirm that nature favors one of these realist theories over quantum mechanics. But suppose it turns out that after many years, or indeed centuries, we have no experimental results which require a modification or completion of quantum mechanics. In particular, what if no limit is found to how large or complex a system can be and still be put in a superposition? Suppose, in other words, that quantum mechanics in its original form appears to be completely correct. Would there be any options for realists?

The reason it is hard to be a realist and believe in quantum mechanics is Rule 2, which gives a special role to measurement. The suddenness of the collapse of the wave function on measurement dictated by Rule 2 means quantum states change in time in a way that pays no heed to locality or energy and instead seems to depend on what we know or believe. Since it makes the quantum state depend on our knowledge, this cannot be part of a realist theory.

Such a theory could not have Rule 2 among its postulates, because that would contradict realism. So we would have to build our theory on Rule 1 alone. This is also a modification of the theory, but it is one it shares with pilot wave theory, so perhaps it's a change worth exploring. Such a theory has no obvious reference to experiment, also no apparent notion of uncertainty or probability, because Rule 1 is deterministic and makes no reference to probability. Can we possibly make such a theory work and stay consistent with realism?

One way to accomplish this would be to derive Rule 2 from a

theory that doesn't postulate it. The collapse of the wave function would happen only in certain special circumstances, such as when an atom interacts with a large, human-size measuring instrument. To do this we have to find roles for uncertainty and probability arising in a world described by a theory that has none.

The project to make sense of quantum mechanics based solely on Rule 1, and in a way that is consistent with realism, has a long history. It was initiated in 1957 by a PhD student of John Wheeler's named Hugh Everett III, and so can be called Everettian quantum mechanics. But it is most often referred to as the Many Worlds Interpretation of quantum mechanics, because some have argued, not without controversy, that it implies that the world we experience is just one of a vast number of parallel universes.

Everett's proposal was presented in his PhD thesis of 1957, and was published the same year.[1] It was unusually short for a PhD thesis, yet it was to have, after a while, a big impact.

Everett, as many have, left academic science just after his PhD to begin a career in the defense industry, so his thesis was his only contribution to physics. And it took many years before it was widely read. But, apart from de Broglie's thesis, I can think of no other PhD thesis which was to have, over the long term, such a disruptive or revolutionary (you choose) effect on the foundations of physics.

One of Everett's ideas is certainly correct and useful. If there is no Rule 2, wave functions don't collapse, so we have to describe what happens in a measurement using only Rule 1. As we saw in our discussion of Schrödinger's cat at the end of chapter 4, interactions, including measurements, lead to correlated states. The example we discussed was

IN BETWEEN = (EXCITED **AND** NO **AND** ALIVE) **OR**
(GROUND **AND** YES **AND** DEAD)

The **OR** signifies a superposition of different possible situations, in each one of which the atom, Geiger counter, and cat are all correlated. Given that they are in a superposition of states, observables such as the aliveness of the cat have no definite value. But Everett noticed that, nonetheless, we can read this superposed state as giving us two contingent statements about the state of the combined system after the measurement. These contingent statements are

If the atom is in the excited state, then the counter will read NO *and the cat will be alive.*

and

If the atom is in the ground state, then the counter will read YES *and the cat will be dead.*

These tell us that the atom, the counter, and the cat have become correlated by the photon's possible passage through the detector.*

The superposed state doesn't tell us which outcome will be observed, but it tells us that the outcome expresses a correlation between the state of the atom and the states of the counter and cat.

This much of Everett's thesis is unimpeachable. It is generally

* Notice that the two contingent statements, which together express the content of the correlated state IN-BETWEEN, do not require or imply that the atom has decayed, releasing a photon that passes through and triggers the detector. At each time, it may have decayed or it may have yet to decay. This is why I refer to "the photon's possible passage through the detector."

true that interactions between two quantum systems set up correlation between the states of the two systems, and these correlations can be read as sets of contingent statements. This is a consequence of Rule 1, applied to interactions.

But notice what this doesn't do. It doesn't tell us which outcome will be observed. Contingent statements may be useful as they give us definite information about the system. But they do not give us complete information. A theory that gave us only contingent statements could not be enough for a realist.

So Everett went further. To make the theory with only Rule 1 realist, he proposed to change our conception of reality. Everett suggested that a state which consists of a superposition of states of detectors describes a reality in which both outcomes happen. In this enlarged reality, both contingent statements will be true. That is, Everett asserted that *a full description of reality is the superposition of the two states.* Part (but only part) of what that implies is that the following statement is true:

> *The atom is in the excited state, the counter reads* NO, *and the cat is alive,* ***and*** *the atom is in the ground state, the counter reads YES, and the cat is dead.*

This would seem to be blatantly false. In the world we live in, the cat experiences only one outcome. This is why in chapter 3 we described the superposition as characterizing an "or." Either she experiences that she is alive, or she is dead and experiences nothing. In our world, it is one or the other.

Everett proposed that the world we experience is only a part of the full reality. In the enlarged world which, he proposed, makes up that full reality, versions of ourselves exist that experience every possible outcome of every quantum experiment.

In other words, the "or" of ordinary experience becomes, in quantum mechanics, an "and." We say "the cat is alive *or* the cat is dead" because the two states are mutually exclusive. But in this formulation, it can nonetheless be true that "the cat is alive *and* the cat is dead."

The idea is that each time an experiment is performed which could have different outcomes, the universe splits into different, parallel worlds, one for each of the possible outcomes. We split as well, along with the world. The experiment creates an additional version of ourselves for each of the possible outcomes. Each version of ourselves lives from then on in a world described consistently by one of the contingent statements we can read off the combined state.

In contrast with pilot wave theory, Everettian quantum mechanics has no particles, so nothing distinguishes the different branches from each other.* We then are invited to regard all branches as equally real, and work out the consequences. So if Everett is right, I am at this moment in Toronto, and I am in London, and indeed simultaneously in myriad places my life might have taken me, including the ocean floor off Peggy's Cove.

These branches are sometimes called worlds. You can see why Everett's proposal has come to be called the Many Worlds Interpretation of quantum mechanics.

For this to work, each version of an observer must have no way to communicate with the others; the branches must be autonomous.

* We can think of Everettian quantum mechanics as pilot wave theory without the particles. In both cases, there is no Rule 2; both make Rule 1 universal. So in both cases, the wave function continually branches, creating alternative histories, such as the ones where I stayed in London or perished off Peggy's Cove with the Swissair flight. The difference is that pilot wave theory has particles, which take only one of the alternate branches.

WHAT I HAVE DESCRIBED so far was Everett's initial version of the Many Worlds Interpretation. On examination, it turned out to be a bit naive, as it ran into several big problems.

The first problem with Everett's formulation is that he suggested that the branching happens when a measurement is made. But this makes measurements appear to be special, whereas it is a basic tenet of realism that measurements are ordinary interactions to be treated like any others.

Indeed, Rule 1 makes no distinction for experiments. So, if you are a realist,* you must insist that what happens for a measurement must happen more generally. The key thing that causes a splitting is an interaction, which produces correlations between the systems that interacted. These correlations can, as we saw, be expressed as contingent statements describing different possible outcomes of that interaction.

To avoid making experiments special, the universe must split each and every time there is an interaction which has more than one possible outcome. But this is happening literally all the time—all that is required is for two atoms to collide with each other, and that is happening myriad times a second just in the air in this room.

Moreover, the interaction that causes the splitting can happen anywhere in the universe. So while you are reading this sentence

* There is an operational reading of Everett that sees the theory purely as a method for producing sets of contingent statements such as I described above, but makes no claims to what is real beyond that. This seems to me a consistent way to read Everett's thesis. (Lee Smolin, "On Quantum Gravity and the Many Worlds Interpretation of Quantum Mechanics," in *Quantum Theory of Gravity: Essays in Honor of the Sixtieth Birthday of Bryce S. DeWitt*, eds. Steven Christensen and Bryce S. DeWitt [Bristol, UK: Adam Hilger, 1984].)

you are splitting a vast number of times, into a vast number of versions of yourself.*

This is a lot to ask someone to believe, all in the name of realism. No wonder it took some time for Everett's ideas to catch on.

A second problem is that if the branching is to replace Rule 2, then it must be irreversible, to reproduce the basic fact that we observers experience every experiment to have a definite outcome. Indeed, the action of Rule 2, which the branching is supposed to replace, is irreversible. But the branching is supposed by Everett to be a consequence only of Rule 1, which is reversible.

A third big problem with giving up Rule 2 has to do with probabilities—or rather, their absence.

Experiments measure probabilities for different outcomes to occur, and comparing these to the predictions of the theory is an important part of testing quantum mechanics. But notice something important: Rule 1 doesn't speak of probabilities. All reference to probabilities in quantum mechanics comes from Rule 2, which gave us a formula for how probable each possible outcome is. That formula, as we noted before, is called Born's rule, and it relates probabilities to the square of the wave function. This is the only part of quantum theory that refers to probabilities, and it is part of Rule 2. If we eliminate Rule 2 from quantum theory, we have nothing left in the theory that speaks of probabilities.

As a result, Everett's version of quantum mechanics tells us only that every possible outcome occurs. Not with some probability, but with certainty.

That is, for every possible outcome of an experiment, the Many Worlds Interpretation asserts there is a branch in which it occurs.

* In the next chapter, we will see that some experts argue that splitting requires a macroscopic process called decoherence. This happens far less often; this has the effect of decreasing "vast" in this sentence to merely very large.

There is no sense in which some branches are more probable than other branches. All Rule 1 can assert is that with certainty, all branches will exist. So we seem to have lost an important part of quantum mechanics—that part which predicts the probabilities that different outcomes occur.

Everett was not dumb; he was aware of this issue, and he attempted to address it. In his thesis, he offered a way to predict probabilities using only Rule 1. To accomplish this, he suggested a way to derive the relation between probabilities and squares of the wave function—a relation which Rule 2 postulates—directly from Rule 1 alone.

At first many were impressed by this result. I know I certainly was when I first read Everett's paper. But it turns out something was concealed in his derivation. Like many erroneous proofs, the argument assumed what was to be proved. The relation between the square of the wave function and probability was snuck into a seemingly innocuous step, which assumed that branches with small wave functions* have small probabilities.† But that was tantamount to assuming a relation between the size of the wave functions and probabilities, and so the proof proved less than was first claimed for it.

Everett's proof did establish one important thing: that if one wanted to introduce quantities called probabilities, it would be *consistent* to assume they follow Born's rule. But it did not prove that it was *necessary* to introduce probabilities, nor did it prove that those probabilities must be related to the size of the wave function.

Yet another problem with Everett's original formulation of the

* That is, wave functions with small amplitudes.

† To put it more precisely, while the *measure* of all those branches with statistics not obeying the Born rule goes to zero, in the limit of an infinite number of trials, the *number* of those branches does not.

Many Worlds Interpretation was that splitting the quantum state into branches is ambiguous. As I explained, each branch is defined by some quantity having a definite value. There is one branch in which the atom is excited and the cat is alive and another branch in which the atom is in the ground state and the cat is dead. But why these and not some other quantities? Ground and excited are states of different energy. But there are other incompatible quantities that we might use instead to define a split. There will be some superposition of ground and excited states that corresponds to the electron being on the left side of the atom and a different superposition that corresponds to the electron being on the right side.

Let's call these states left and right. Why not split with respect to these? These would lead to states of the cat which were superpositions of alive and dead. She no longer would experience a world where there are definite outcomes to experiments. But Rule 1 doesn't care whether she experiences definite outcomes or does not. We call this the preferred splitting problem.

At first there seems to be an obvious answer to the preferred splitting problem: we must split the wave function so that the different branches describe situations in which macroscopic observers like the cat see definite outcomes.

But this is tantamount to reintroducing Rule 2, because it gives what macroscopic observers see a special role. You have not solved the mystery of why macroscopic observers see definite outcomes. And by giving observers a special role, you give up on achieving a realist interpretation, which must be based on hypotheses about what is real in the absence of observers.

ELEVEN

Critical Realism

There is a near consensus among the people who have examined the original version of the Many Worlds Interpretation, the version put forward by Everett and championed by Wheeler and DeWitt, that it fails as a realist approach to quantum theory. Either you make measurement special and give up on realism, or you face the big issues I raised. The most important of these are the preferred splitting problem and the question of where the theory contains the probabilities, and the related uncertainties, that experimentalists measure.

So, can the project of giving a realist version of quantum theory, based only on the wave function evolving strictly according to Rule 1, be saved?

In recent years some rather radical solutions have been offered to the two big puzzles—the preferred splitting problem and the question of where the probabilities come from. The preferred splitting problem is widely thought to have been solved by an idea called *decoherence*, which I will explain shortly. Ideas about the origin of probabilities mostly originated from a group of deep thinkers at Oxford, centered in its philosophy department. The new approach to probabilities was formulated by David Deutsch, and it has been extensively studied and developed by his Oxford colleagues.[1]

Oxford has had a very smart group of philosophers of physics, and several of them have focused on making sense of Everett's ideas. They have included Hilary Greaves, Wayne Myrvold, Simon Saunders, and David Wallace.* Together with Deutsch and a few others, they have put forward what has sometimes been called the Oxford interpretation of quantum mechanics.[2] These proposals and the arguments offered in their support are both ingenious and subtle, but so are the objections made by several physicists and philosophers. Given the very high level of careful thought that has gone into these developments, I think it is fitting to call this an episode of critical realism.

After many spirited and elaborate arguments, the project of making sense of a realist quantum theory based only on Rule 1 is still in progress. The issues are surprisingly intricate and elusive, and there is as yet no general agreement among experts as to what has been achieved. To make it even more complicated, the proponents disagree among themselves, so that among the five or six main initiators of this view, several different versions are defended, which differ in subtle but important ways from each other. Consequently, I can present only a rough introduction to the key ideas and issues behind this new "Oxford interpretation."

* Wallace and Myrvold have since left Oxford; Deutsch, Greaves, and Saunders remain as of 2018.

The idea of decoherence starts with the observation that a macroscopic system, such as a detector or an observer, is never isolated. Instead, it lives in constant interaction with its environment. The environment is made up of a vast number of atoms all moving about unpredictably hither and thither, and this introduces a big dose of randomness into the system. This random element affects the motions of the atoms which make up the detector. This, roughly speaking, leads the detector to lose its delicate quantum properties and behave as if it were described by the laws of classical physics.

Consider what an observer can learn by looking at a detector. The observer is also a big object made of vast numbers of atoms, all in contact with a random environment. If we look at the detailed small-scale behavior of the atoms making up the detector and the observers, we will see chaos, as the picture will be dominated by the random motions of the individual atoms, both our atoms and those in the detector. To see any kind of coherent behavior we have to look at bulk, large-scale motions of relatively large pieces of the detector. These require averaging over the motions of myriad atoms. What emerges are bulk quantities which measure macroscopic quantities such as the color of a pixel or the position of a dial. Only these behave reliably and predictably.

Indeed, these bulk quantities behave as if the laws of Newtonian physics are true. It is only when we focus on these bulk, large-scale quantities that we can perceive something irreversible to have happened, such as the recording of an image, in which each pixel comprises a vast number of atoms. And according to this picture, it is only when something irreversible has happened that we can say that a measurement has taken place.

Decoherence is the name we give to the process by means of

which irreversible changes emerge by averaging out the random chaos of the atomic real. Decoherence is a very important feature of quantum theory, for it is why the bulk properties of large-scale objects, such as the rough motions of soccer balls, swing bridges, rocket ships, planets, and so forth, appear to have well-defined values, which obey the laws of Newtonian physics.

The word "decoherence" refers to the fact that such bulk objects appear to have lost their wave properties, and so they behave as if they are simply made of particles. According to quantum mechanics, everything, including cats, soccer balls, and planets, has wave as well as particle properties. But for these bulk objects, the wave properties have been so randomized by their interactions with their chaotic environment that they cannot be accessed in any experiment, so the wave half of the wave-particle duality has been rendered mute, and the objects behave like ordinary particles.

But sometimes there is more than one way the system could decohere. A perfect example of this is Schrödinger's cat. The cat could decohere as a live cat or it could decohere as a dead cat. What makes the difference is a quantum variable: if the atom was decayed the cat would decohere as dead; if the atom was excited the cat would decohere as living. So a detector is a kind of amplifier, with a filter that only allows it to register states where the atom is definitely either excited or decayed.

The puzzle, you may recall, was: What happened to the cat while the atom existed in a superposition of excited and decayed? The answer is still the same—if you look at the quantum state microscopically: it is a correlated superposition of an excited atom with a live cat superposed with a decayed atom and a dead cat.

But if you look only at bulk properties so that decoherence can do its work, the randomness turns the superposition into an almost irreversible change. Now there are two outcomes—the live cat and the dead cat—and both emerge! This, according to the decoherence story, is how the world splits in two.

The Oxford thinkers then claim that the branchings and the splittings of the wave function are defined by decoherence.* The split is made so that it separates different outcomes which have different values of macroscopic properties, such as the position of a dial.

The main claim then is that only those subsystems which decohere can be counted on to have observers associated with them. As we are interested in what observers see, we should focus on these and throw the rest away. This opens up a route to deriving probabilities in which you compare only the likelihoods of what would be observed on branches that decohered.

This introduces a notion of observers into the theory, which might be thought to weaken its claim to realism. However, this is a way of discovering a role for observers that arises from the dynamics of the theory, which is surely better than just postulating a special role for observers at the beginning. One might argue that probabilities are not intrinsic to the world, but are only aspects of observers' beliefs about the world. Then such a description could be consistent with realism because there is an objective characterization of a property that distinguishes observers from other subsystems. Observers are subsystems that decohere.

Decoherence solves the preferred splitting problem because decoherence takes place only with respect to certain observables. Often these are positions of large-scale objects.

Before we go further, I should mention that there is, unfortunately, a problem with making decoherence a necessary part of the interpretation of the theory, which was pointed out a long time ago

* The idea that decoherence defines the branches in the Many Worlds Interpretation had been suggested earlier by others including Heinz-Dieter Zeh, Wojciech Zurek, Murray Gell-Mann, and James Hartle.

by my teacher Abner Shimony. This problem can be put very simply. Rule 1 is reversible in time, so every change a state undergoes under Rule 1 can be undone, and indeed will be undone if we wait long enough. But Rule 2 is irreversible, and the way it introduces probabilities for the outcomes of measurements makes sense only if measurements are irreversible and cannot be undone. Thus, Shimony argued, it is impossible that Rule 2 could be derived from Rule 1 alone.

As I described it above, decoherence is an irreversible process in which coherence of states, needed to define superpositions, is lost to random processes in the environment of the measuring instrument. But how can decoherence arise in a theory based on Rule 1 alone, as all changes dictated by Rule 1 are reversible in time?

The answer is that decoherence is always an approximate notion. Complete decoherence is impossible. Indeed, if we wait a very long time, decoherence will always be reversed, as the information needed to define superpositions seeps back into the system from the environment.

This is due to a general theorem, called the *quantum Poincaré recurrence theorem.*[3] Under certain conditions, which can be expected to hold for systems containing an atomic system plus a detector, there is a time within which the quantum state of the system will return arbitrarily close to its initial state. This time, called the Poincaré recurrence time, can be very large, but it is always finite. The conditions include that the spectrum of energies be discrete, which is certainly reasonable.*

* Indeed, the holographic principle (which is defined on pages 259–260) requires that any system that can be fit into a box with walls of a finite area has a finite number of states. This is certainly the case with any system of the kind we are discussing here—an atomic system interacting with a measuring instrument. One reply that might be given is that we live in an ever-expanding universe, which may imply that the dimension of the state space is continually expanding, in which case there is no Poincaré recurrence. This raises several fascinating issues. But, for the moment, it is

Decoherence is a statistical process, similar to the random motion of atoms that leads to increases of entropy, bringing systems to equilibrium. These processes appear to be irreversible. But they are actually reversible, because every process governed by Rule 1 is reversible. This is true in both Newtonian and quantum physics; both have a recurrence time. In either case the second law of thermodynamics, according to which entropy probably increases, can hold only for times much shorter than the Poincaré recurrence time. If we wait long enough, we will see entropy go down as often as it goes up.

Similarly, one might try to argue that over shorter times, there is a low probability for decoherence to reverse, giving way to recoherence.

Now, as long as we are interested only in what happens over much shorter times than it takes to recohere, and we want only an approximate description of what goes on when atomic systems interact with large bodies, suitable for practical purposes, decoherence provides a useful approximate description of what happens during a measurement. Indeed, decoherence is a very useful concept when analyzing real quantum systems; for example, much of the design of a quantum computer goes into counteracting decoherence. But as a matter of principle, that description is incomplete, as it leaves out the processes that will recohere the state if we wait long enough.

However, when the state recoheres, measurements based on decoherence are undone. Therefore, measurements as described by Rule 2 cannot be the result of decoherence, at least as decoherence is described in a theory based only on Rule 1.

So it seems that decoherence cannot alone be the key to how

---

enough to note that this amounts to claiming that quantum mechanics makes sense only when applied to the universe as a whole.

probabilities appear in the Everett quantum theory, because it is based solely on Rule 1.

THIS DISCUSSION MAKES it clear that the question of where probabilities come from is central to making sense of the Many Worlds Interpretation. The key to understanding the Oxford approach lies in understanding what a probability is. This question is far more difficult than it appears. We all have an intuitive idea of what it means to say the probability of a flipped coin landing on heads is 50 percent. People know the difference between what to expect when the forecast says the chance of rain tomorrow is 10 percent and the forecast says the chance of rain is 90 percent. But when we look into what we actually mean when we talk of probabilities, we find the notion gets surprisingly slippery.

Part of the reason probability is confusing is that there are at least three different kinds, or meanings, of probability.

The simplest notion is that probability is a measure of our credence or belief that something will happen. When we say there is a 50 percent chance of heads on the next coin toss, that is not a statement about the coin; it is a description of our belief about the result of tossing the coin. These are called *Bayesian probabilities*.

When we say the Bayesian probability for rain tomorrow is 0 percent, that is just a way of saying we believe it will not rain, and when we say that probability is 100 percent, that says we are sure it will. Probabilities between them, such as 20 percent, 50 percent, or 70 percent, refer to the strength of our belief that it will rain. In particular, when we say something has a 50 percent probability of happening, we are really confessing we have no idea whether it will happen.

Bayesian probabilities are clearly subjective. They are best evaluated in terms of our behavior. The higher the probability for rain, the more likely it will be that we would bet on rain, or at least carry an umbrella.

Many probabilities we deal with in ordinary life are best understood in this way, as betting odds. Certainly, probabilistic predictions about the stock market or the housing market are of this kind. Indeed, most of the time when we refer to the probability of some future event, we are making a subjective statement of belief, using Bayesian probabilities.

A second kind of probability comes into play when we keep records of the relevant events. If we toss a large number of coins and keep records of how often they come up heads, we can define the proportion of heads in that sequence of tosses to be a probability. These are called *frequency probabilities*.

Batting averages and other sports statistics are frequency probabilities. They give the proportion of the times that a batter got on base after he was at bat.

Sometimes weather forecasts are of this kind. When the National Weather Service website tells us in the morning that there is a 70 percent probability that it will rain this afternoon, what they might be saying is that within their vast records, roughly 70 out of 100 days with conditions like those of this morning had rain in the afternoon.

Of course, these probabilities are imprecise. The problem with these is that so long as the number of days observed is finite, the frequencies will vary. But the more days of which the weather service has records, the more reliable the forecast will be.

If one flips a coin 100 times, then one can ask how often one gets heads. The proportion is called the relative frequency of getting heads. This will tend to be around 50; we are not surprised if it turns out often to be 48 or 53.

For any finite number of trials, then, the number of heads will rarely be exactly half. The key idea is that, were we able to do an infinite number of trials, the proportion of different outcomes would tend to some fixed values. This defines the relative frequency notion of probability.

The problem with this is that in the real world, we only get a finite number of tries. As long as the number of trials is finite, there is a good chance that the number of heads will be different from exactly half the trials. A surprisingly hard question to answer is what it takes to show that a probabilistic prediction is wrong, given that we can only do a finite number of tests. Indeed, often all we can say is that our prediction is improbable. But for this to be meaningful we have to define what we mean by improbable. We cannot assume we know what improbable means as we are in the process of defining it.

Suppose we toss a coin a million times and come up with 900,000 heads. It is possible that this is a rare fluke and our coin is normal. But we can conclude that it's very probable—although not certain—that the coin is weighted.

By definition, we choose our subjective probabilities. But we can ask that there be a relation between the subjective Bayesian probabilities we choose and objective frequencies taken from past records. So long as we have no more information, the best bet we can make is the one that follows the odds that are based on historical records. What we mean here by "best bet" is the choice that, most of the time, will serve our interests. In economic-speak we could say that this is the "most rational choice."

We might put this as follows:

It is most rational, in a situation where you have limited knowledge, to choose to align your subjective betting odds with the frequencies observed in the historical record.

This is a version of the "principal principle" of the philosopher David Lewis. This principle has at its root an assumption that, everything else being equal, the future will resemble the past. Or at least that, given incomplete information, it is rational to bet on the future resembling the past. This bet may sometimes put you on the wrong side of history, but it is still the safest bet you can make.*

Now, suppose we ask a different question, which is to explain the frequencies observed in the records of a particular experiment. Suppose the frequency observed was close to 50 percent. It would be natural to try to explain that result by an application of the laws of physics to the particular experiment.

Such an explanation might give reasons why heads would be as likely an outcome as tails. This would include the hypothesis that the coin was fair, as well as hypotheses about the tosses, how the coin behaves when it hits a surface, and so on. Our explanation might also refer to results from other experiments, which support our belief in the theory.

Once we have such an explanation, we would use it to predict that a single toss has an equal chance to end up heads or tails. This prediction is a belief, and hence a subjective Bayesian probability. But it refers to the single toss. This toss need not be part of a large number of trials; hence no relative frequency is involved. It then makes sense to say that the particular coin has, in its context, a

* Note that this principle cannot be taken as an end of the story of making sense of probabilities, because it isolates something that we would really like to understand from first principles. What is missing is a convincing argument that would compel us to line up our subjective probabilities with the objective chances.

physical *propensity* for a single throw to end up heads 50 percent of the time.

The propensity is an intrinsic property the coin has as a consequence of the laws of physics. It can be expressed as a probability, but it is not a belief. Rather, it justifies a belief. It is something in the world that we may have a belief about. Nor, as we said, is a propensity a frequency, for it is a property of the coin, which applies to each individual toss. Propensity would then seem to be a third kind of probability, different from either beliefs or frequencies.

Note that unlike the other two kinds of probabilities, propensities are consequences of theories and hypotheses about nature. But they have distinct relations to the two other kinds of probabilities. We can have beliefs about propensities. Propensities in turn can explain relative frequencies and can justify beliefs.

In ordinary quantum mechanics, probabilities arise from Rule 2, in particular the Born rule, which connects the probability of seeing a particle at some position to the square of the amplitude of the wave at that position. That probability is posited to be an intrinsic property of the quantum state; hence it is a *propensity probability*. Quantum mechanics asserts that there is no deeper explanation for that probability and the resulting uncertainty; it is an intrinsic property of the quantum state.

When Everett dropped Rule 2, the result was a theory without any notion of probability, intrinsic or otherwise. As I described, he tried and failed to replace this with a frequency notion of probability.

The dilemma proponents of the Everett formulation of quantum mechanics faced was that there are branches in which observers see that Born's rule connecting magnitudes with frequencies holds, and there are other branches whose observers see that Born's rule is violated. Let's call these benevolent branches and malevolent branches.

The latter may have smaller wave functions than the former, but one cannot use this to argue that the latter are any less probable, because to do so would be to impose on the theory the relation between size or magnitude of the wave function and probability. But that is exactly what proponents of Everett's formulation are trying to derive from Rule 1; to assume it would be to sneak in Rule 2 by the back door.

THE EVERETT THEORY is a hypothesis about the nature of reality. It posits that all that exists is a wave function evolving deterministically. From the imaginary perspective of a godlike observer outside the universe, there are no probabilities, because the theory is deterministic. All branches of the wave function exist; all are equally real.

The Everett theory asserts that each of us leads many parallel lives, each defined by a branch that has decohered. The theory also tells us that each of these branches exists, with certainty. So if this theory is right, since there is no Rule 2, there are no objective probabilities at all. Let us call this Everett's hypothesis.

But we are not godlike; we are observers living inside the universe, and, according to the hypothesis, we are part of the world that the wave function describes. So that external description has no relevance for us or for the observations we make.

We are then faced with a puzzle. Where in this world do we find the probabilities that ordinary quantum mechanics claims to predict, which are to be compared with frequencies counted by experimentalists? With no Rule 2, these probabilities are not part of the world as it would be in our absence. Frequencies are counts of definite outcomes, but such things are not unique or exclusive facts in Everettian quantum theory, because given any possible counting of outcomes of a repeated experiment, there are branches which have that count. There are branches in which those counts agree with the predictions of quantum mechanics (with Rule 2) and branches in which they don't. We cannot say the former are more probable than the latter, because in Everettian quantum theory there are no objective probabilities. We cannot even say that the former are more numerous than the latter because in realistic cases there will be infinite numbers of each.

You read this right: Everettian quantum mechanics predicts that an infinite number of observers will observe experimental results that disagree with the predictions of quantum mechanics! That is the fate of the infinite number of observers whose ill fortune takes them along malevolent branches. It is also the case that an infinite number of observers on benevolent branches see experimental results consistent with quantum theory's predictions. But that is small consolation, because a benevolent branch can turn malevolent at any moment.

What it seems we cannot say, in Everettian quantum mechanics, is that quantum theory predicts objective probabilities, which are inherent features of nature that exist in our absence. And, unless we find another way to introduce probabilities, we cannot say that the theory can be tested by doing the experiment and counting the different outcomes, because the failure of any such test can be dismissed by supposing that we are just on a malevolent branch—and those are not any less probable or any less numerous than the benevolent branches which confirm the probabilistic predictions of quantum mechanics.

To address this situation, David Deutsch made an interesting proposal, which was to ask not whether the Everett theory is true or false, but how we, as observers inside the universe, should bet, were we to assume that it is true. In particular, the major thing we have to bet on, assuming the Everett story is true, is whether the branch we live on is benevolent or malevolent. Every other bet we might make depends on that single bet. If we are on a benevolent branch, then bets we place based on Born's rule will pay out. If we aren't so fortunate, then all bets are off, because literally, anything could happen.

This is not a bet about the universe, because it is certain that the universe contains observers who live on both kinds of branches. It is instead a bet on where we are in the universe. There is no right answer to this question, because, if Everett is right, there are both kinds of observers, and some of us will be one kind, some the other.

Nonetheless, Deutsch proposes that it is more rational to bet we are on a benevolent branch. The argument is technical and employs a branch of probability theory called *decision theory*. Deutsch's result then assumes certain axioms of decision theory, which specify what it means to make a rational decision.

Some experts have criticized this approach; some defend and develop it, while still other experts offer alternative arguments to the same conclusion. Given that I am not a specialist in this area, I am not going to speculate on which experts are right.

But notice what this kind of argument doesn't—indeed cannot—do. It cannot offer us evidence that the Everett hypothesis is true, because Deutsch and his colleagues begin by assuming that the hypothesis is correct. Their arguments also assume the axioms of decision theory. If you don't accept them you do not prove that the probabilities are related to the magnitudes. All the argument could show is that, assuming the axioms of decision theory, it is consistent with the Everett hypothesis to place bets, and make other kinds of decisions, as if Born's rule were true.

Notice that, even given the assumption that Everett is true, the observers modeled as part of an Everett world do not know that they live in an Everett world. There is no reason they should, and if they nonetheless did, they would not be models of us, as observers in a universe whose full set of principles remain to be discovered. For them as for us, the Everett hypothesis must be one of several competing hypotheses as to the nature of the beables of the quantum universe.

Let us then consider the situation of observers inside an Everettian universe. There are two cases, depending on which kind of branch we live on. Suppose we are fortunate and live on a benevolent branch, so that our bets based on Born's rule pay out. Well then, by definition, we do no better, and no worse, than people who believe in other formulations and interpretations of quantum mechanics and so also place bets based on Born's rule. What the other approaches lack is a justification based on decision theory. On the other hand, pilot wave theory and collapse models have no need of such justification, because they rely on completely objective notions of probability arising from our ignorance of the details of the individual experiment.

Thus, on its own terms, in which it cannot address what is true, but can only offer advice about how best to place bets, Deutsch's argument implies that it is no more rational for observers inside an Everettian world to believe in Everett than it is for them to believe in Bohr or de Broglie, Bohm or any other interpretation. So, in the best case, even assuming that Everett is right, observers in an Everettian world cannot muster any evidence to believe Everett's hypotheses over the alternative hypotheses.

What about the versions of ourselves that live on malevolent branches? Their bets based on Born's rule don't pay off because the frequencies they measure disagree with those predicted by Born's rule. So how does it look from the point of view of these unlucky observers? Remember that for them, the usual formulation of quantum mechanics (say, as presented in von Neumann's book) must be a hypothesis, and Everett's hypothesis is a different, competing, hypothesis.

Observers on a malevolent branch conclude that the first hypothesis is simply false because Born's rule does not predict the results they observe. The second hypothesis, Everett's, is not falsified, because that predicts that some observers will see Born's rule fail. But it's worse than that. Given any results of repeated measurements, Everett's story predicts that some observers, living on a malevolent branch, will see exactly those results. So Everett's hypothesis cannot be falsified by testing any probabilistic prediction based on Born's rule, as there is no outcome of a repeated measurement that is inconsistent with an Everettian universe.

So it seems that the bulk of experimental predictions that could falsify ordinary quantum mechanics—those that compare theoretical probabilities to experimentally observed frequencies—would not count as falsifying Everettian quantum mechanics. While not completely unfalsifiable—because the theory makes other kinds of predictions, which do not involve probabilities—Everettian quantum mechanics seems to be far less vulnerable to falsification than ordinary quantum mechanics.

That, in itself, is a good reason to prefer an alternative approach. A theory that is less falsifiable is by definition less explanatory.

On the other hand, if we accept the assumptions of Deutsch and the other Oxfordians, then we must disregard the point of view of the malevolent branches, because those branches are very improbable. In this case there is work that shows that once one neglects the malevolent branches, the theory is testable.

The Oxfordians emphasize that if you assume the axioms of decision theory are correct, then you are allowed to deduce that it is rational to reason as if the magnitudes are related to the probabilities. It follows that it is rational to reason as if we have a very small probability of ending up on a malevolent branch, so that possibility can be ignored.

They might further claim that something like this is always the case when we reason probabilistically. We could always be unlucky and have a coin toss result in heads a thousand times in a row. But there is a difference. In a finite life, in a single finite world, we can rest assured that such things almost never happen. But in strong contrast, Everettian quantum mechanics asserts that correspondingly malevolent branches not only exist—they are as numerous as benevolent branches. While Deutsch's argument tells us about subjective betting probabilities taken by observers inside the Everett world, it remains the case that the overall theory is deterministic and that each of the branches definitely exists.

It seems, at least as best I've been able to understand, that the attempts by Deutsch and others* to rescue the project of making sense of the Everett hypothesis by means only of subjective probabilities for observers in an Everett universe, introduced via decision theory, do not convincingly succeed. Arguments based on subjective notions of probability alone fail to explain why we can neglect the malevolent branches—for, if Everett is right, they are objectively real.

* Including Greaves, Myrvold, and Wallace. I should note that they introduce lines of argument I haven't mentioned here, about which experts disagree, so the situation is somewhat more complex than the overview I have presented.

SOMETHING NEW IS NEEDED. To save the day, Simon Saunders has proposed to cut the Gordian knot by positing that the magnitudes of the branches give objective probabilities (rather than betting probabilities) of an observer finding themselves on a decohered branch, in agreement with Born's rule. His argument for this is that the magnitudes of the branches do indeed turn out to have many of the properties we would want objective probabilities to enjoy. Indeed, his claim is that they have these properties as a consequence of Rule 1—hence this is a discovery of a consequence of the laws by which quantum states evolve. It is not an additional postulate, as Rule 2 is. If his proposal succeeds, it would be a genuine derivation of Rule 2 and the Born rule from the theory based purely on Rule 1.

This gets us out of the problems raised by the malevolent branches, because, assuming Saunders is right, it is not very probable to find ourselves on one of them. But Saunders claims it would accomplish more than that: it would be a genuine derivation of how objective probabilities arise in nature, and it would explain why we must align our subjective betting odds with the objective probabilities.

My understanding is that the experts in Oxford are at present divided as to whether Saunders's proposal succeeds. One issue is that the branch magnitudes have some but not all properties of objective probabilities. So we must leave this discussion here; after more than sixty years of study, it is still unresolved whether sense can be made of Everett's startling idea.

RECOGNIZING THAT THE PROJECT of making sense of Everett's hypothesis remains a work in progress, I can offer a series of remarks.

My overall understanding is that the Everett hypothesis, if successful, would explain vastly too much, and also much too little. Too much, because we have to believe that the whole world we used to think of as real is just one branch within a vastly larger reality. And too little, because a great deal is left out of this picture of reality. What is most characteristic about experienced reality is that every process we observe has a definite outcome. What is also most impressive about quantum theory is its ability, using Rule 2, to make precise predictions of the observed frequencies of those definite outcomes. What I want from realism is a detailed explanation for how those probabilities arise as relative frequencies, by averaging over a set of repeated runs of the experiment.

The reality that we realists seek is the world as it is, or would be, in our absence. Subjective probabilities that guide decision makers to place bets are not part of that world, since they would not exist if we did not. The question is not whether decision makers are real, for we are certainly real. Nor is the question whether we could, if we were interested, seek a scientific account of what constitutes rational decision making. The question is instead whether we can realize the ambition of physics to describe light and atoms in a way that is completely independent of whether we exist or not.

Let me emphasize that the jury is still out as to whether the Oxford approach succeeds, on its own terms, in making sense of the Many Worlds Interpretation. The Everett hypothesis may yet be shown to be inconsistent or incoherent. Or it may turn out to be the only realist approach to quantum mechanics, in which Rule 1 alone suffices to frame the theory. For me, either outcome would just strengthen the argument that we need a new theory.

When empirical tests of theories fail, we still have to make decisions about which theory to work on. As a number of philosophers and historians have stressed, before definitive evidence is in, there is no avoiding bringing in factors that may seem nonscientific when evaluating which research program and theory is deserving of our time and attention. This is especially the case because these are in part individual decisions, and when empirical criteria have yet to be decisive, it is in the interests of the scientific community as a whole to encourage the widest diversity of approaches consistent with the evidence in hand at the moment. As Paul Feyerabend explains in his book *Against Method,* it is competition among diverse viewpoints and research programs that drives the progress of science, especially through critical periods when the evidence is not sufficient to decide which approach will ultimately yield the best explanations.

Evaluating a research program based on non-empirical factors is partly a matter of individual taste and judgment.* After a lot of effort to understand the thinking of its proponents, here is how the case for the Everett program seems to me. I expect, indeed I know, that others who have thought it through do not agree. I am not afraid to confess that no issue in quantum foundations has been more challenging and more painful to me personally than the issue of Everett, where I find myself in disagreement with friends and colleagues for whom over the years I have grown to have great respect.

* My own view as to how science works and the role of individual judgments in forming a consensus of the whole scientific community is outlined in chapter 17 of my book *The Trouble with Physics.*

We know that the original form of the Many Worlds Interpretation fails as a realist approach because it runs into two big problems, which are the preferred splitting problem and the problem that the theory is deterministic and has no probabilities. After a great deal of effort to develop a more sophisticated version based on decoherence and subjective probabilities, experts continue to disagree over technical issues. But even if they do succeed, what would be established is that the axioms of decision theory require that observers living in an Everett world bet as if Born's rule were true. That does not, however, give us a reason to believe that we live in an Everettian universe. Nor am I aware of any empirically based argument that would require us to prefer Everett over other approaches. Despite some provocative claims to the contrary, there is no experimental outcome that cannot be explained at least as well by the other realist approaches. There are claims that Everett alone can explain phenomena such as the speed-up of quantum computing, but these are contradicted by the fact that the alternative realist programs, such as pilot wave theory, provide accounts of these experiments which are at least equally explanatory.

One argument for Everett begins with the assertion that there are only three realistic formulations of quantum theory, and that the other two, pilot wave theory and collapse theory, have tensions with relativity and hence have trouble incorporating quantum field theory. This argument then implies that, assuming it can be made sense of, Everett must be correct. I disagree, and take this as strong motivation to seek to invent other realist approaches, as I describe in the closing chapters of this book.

That is where the scientific case leaves off; let's then turn to non-empirical factors. The philosopher Imre Lakatos recommended investing in research programs that are progressive, by which he meant that they are rapidly developing and have the potential to lead to a breakthrough. A progressive research program is also one that is open to future developments and surprises, in contrast to programs which assume the basic principles and phenomena are understood. Progressive criteria favor realistic approaches to quantum foundations over the anti-realist approaches because the latter confine us to developing new ways of talking about quantum phenomena which are assumed to be already known, while the former understand that quantum mechanics is incomplete and hence aim to discover new phenomena and new principles in which to situate them.

Within the realist approaches, I believe there is a case to be made that Everett's hypothesis is the least progressive—although there are arguments on both sides. An enormous effort has gone into developing Everett quantum mechanics, much of it technical and extremely clever, but most of that work has gone to addressing problems that arise only in the Many Worlds Interpretation, but do not trouble the other approaches. I might suggest that the Everett program is, of the realist approaches, the least open to the possibility that future discoveries will lead us to modify the principles and the mathematical formalism of quantum mechanics.

On the other side, it should be pointed out that the Everett theory stimulated much work on decoherence, which was important generally for our understanding of quantum physics. It also inspired, and continues to provoke, much progress in quantum computing. The Many Worlds Interpretation played a role in the pioneering work of David Deutsch. Yet we must also credit pilot wave theory and collapse models for the experimental proposals they have stimulated regarding, for example, out-of-equilibrium physics in the early universe. So it seems that an argument about which realist approach is more progressive comes out about even.

The odd thing about the Oxford approach is that, while it tells us nothing about the world we experience that we didn't already know, or couldn't have deduced within other versions of quantum theory, it has a lot to say about all the worlds we don't and can't experience, and especially about the near copies of ourselves which populate them. Given that they are presumably just as alive and just as conscious as we are, I find myself wondering if we—or those of us who believe in Everett enough to contemplate it as a serious possibility—should care about our copies, and whether we have any responsibilities toward them.

I admit that to inquire into the quality of lives of our copies on other branches may seem a bit academic. But one thing we academics are trained to do is to work out the logical consequences of hypotheses and assumptions. And the most provocative and, to me, distasteful consequence of Everett is that we must believe that each of us has an infinite number of copies, each every bit as alive and conscious as we are. This sounds more like science fiction than science, but it does seem a straightforward consequence of Everett's hypothesis. Since this is science and not faith, we don't have the option of taking a "liberal" interpretation of Everett in which we choose to believe certain aspects, such as the existence of a wave function of the universe, while ignoring others.

It then seems to me that Everett raises two kinds of ethical quandaries. First, it condemns a vast number of living and conscious beings to suffering which cannot be mitigated by efforts we make. Beyond that, I would worry that the fact that many of our most talented and accomplished scientists believe we live in that unhappy universe is inimical for the long-term public good, because, by erasing the distinction between possibility and actuality, it diminishes the motivation to work to improve our world.

Couldn't we say the same about the increase in entropy mandated by the second law of thermodynamics, which is ultimately the cause of the death of most living creatures? The difference is that we know the second law is true. We have no choice whether to believe it, whereas there are alternative formulations of quantum theory which do not impose on us the existence of copies. It is also very legitimate to criticize the scientists and philosophers who drew unnecessarily pessimistic conclusions based on an incomplete picture that neglected the positive effects of self-organization in far-from-equilibrium systems.

The whole notion of an observer "living" on a malevolent branch can be objected to on the grounds that none of the biochemistry that life depends on would function well in a world in which Born's rule regularly failed. To be more precise, we might rate malevolent branches by the proportion of events in which Born's rule failed to hold. We could then catalogue branches by the severity of such failures. Living in a mildly malevolent branch would be like being subject to a low dose of ionizing radiation, with similar consequences of decreased health.

Even among the benevolent branches there would be disparities in health. Tomorrow a gamma ray will strike a strand of my DNA, and the consequences will include the splitting of us and our world into a bunch of decohering worlds. Some of my copies will develop cancer as a result; some won't. There are versions of me in both sets; hence I care about both. The extreme version of this argument suggests that, far into the future, some very fortunate copies of me, who had the luck to dodge every bullet and survive every cancer, will be still alive.

It seems to me that the Many Worlds Interpretation offers a profound challenge to our moral thinking because it erases the distinction between the possible and the actual. For me, the reason to strive to make a better world is that we can hope to make the actual future better than the possible futures we were dealt to begin with. If every eventuality we worked to eliminate, whether starvation, disease, or tyranny, was actual somewhere else in the wave function, then our efforts would not result in an overall improvement. Issues such as nuclear war and climate change are less urgent if there are multiple versions of Earth and the human race has more than one chance to get things right.

The existence of all these copies of ourselves would then seem to me to present a moral and ethical quandary. If no matter what choices I make in life, there will be a version of me that will take the opposite choice, then why does it matter what I choose? There will be a branch in the multiverse for every option I might have chosen. There are branches in which I become as evil as Stalin and Hitler and there are branches where I am loved as a successor to Gandhi. I might as well be selfish and make the choices that benefit me. Irrespective of what I choose, the kind and generous choice will be made by an infinite number of copies living in an infinite number of other branches.

This seems to me to be an ethical problem because simply believing in the existence of all these copies lessens my own sense of moral responsibility.

A dear friend who works on Everettian quantum theory would insist that, nonetheless, this is a way the world might be. Our job is to figure out how the world is, and it is not up to us to impose our personal likes and dislikes. My reply is that, so long as there is no decisive argument to prefer Everett over other approaches, I am free to bet on another approach. They are free to do otherwise, but

I choose to invest my time in developing cosmologies that inspire us to look for new particles, new phenomena, new physics, over the scholastic contemplation of the lives of copies of ourselves.

And, I might add, given that I don't believe it is likely that Everett or anything like it is going to turn out to be true, there is little danger of harm if a few brilliant philosophers choose to spend their efforts working out the consequences of a truly startling and subtle hypothesis. (Were the idea to come to influence the zeitgeist, that would be something else to worry about.) Even if it is a wrong idea, it is an idea that probably had to come up sooner or later, and they have the right kind of analytically able minds, rigorously trained, that are suited to the question, which mine is clearly not. Let us then hope they will finally resolve the question of whether or not a realist theory based on Rule 1 alone can make sense.

THE DISTINGUISHED PARTICLE THEORIST Steven Weinberg recently weighed in on the failure of efforts to deduce probabilities from quantum mechanics.

> There is another thing that is unsatisfactory about the [Many Worlds] realist approach, beyond our parochial preferences [e.g., "not liking" the idea of having copies]. In this approach, the wave function of the multiverse evolves deterministically. We can still talk of probabilities as the fractions of the time that various possible results are found when measurements are performed many times in any one history; but the rules that govern what probabilities are observed would have to follow from the deterministic evolution of the whole multiverse. . . . Several attempts following the realist approach have come

> close to deducing rules like the [probability] Born rule that we know work well experimentally, but I think without final success.[4]

THERE IS A LAST MORAL to draw from the story of Everettian quantum mechanics. Some of its proponents claim that Everettian quantum mechanics *is* quantum mechanics, and that all else is a modification of it. But that is simply not the case. Ordinary textbook quantum mechanics—by which I mean the theory that is taught in the standard textbooks (Dirac, Bohm, Baym, Shankar, Schiff, etc.), and therefore the theory in common use by real physicists—is based on Rule 1 *and* Rule 2. That theory simply does not have a realist interpretation.

So realism, in any version, has a price. The question is only what price we have to pay to get a new theory that makes complete sense and describes nature correctly and completely.

PART 3

BEYOND THE QUANTUM

TWELVE

Alternatives to Revolution

In the end we are driven to search for what we hope will turn out to be the correct ontology of the world. After all, it is the desire to understand what reality is like that burns deepest in the soul of any true physicist.

—LUCIEN HARDY

In the last few years the field of quantum foundations has undergone a lively ascension. After eight decades in the shadows, it is finally possible to make a good career as a specialist in quantum foundations. That is for the good; however, most of the progress (and most of the young people) has been on the anti-realist side of the field. The aim of most of the new work has not been to modify or complete quantum theory, but only to give us a new way of speaking about it. To explain why, I need to review a bit of the history of the field of quantum foundations.

Quantum mechanics did not spring up overnight. It was the result of a long gestation, which began in 1900 with Planck's discovery that energy carried by light came in discrete packets, and culminated in the final form of quantum mechanics being established in 1927. There followed a period of debate among the founders, during which many of the quantum physicists were concerned with the foundations of the new theory. However, this period of free debate soon came to an end, and, despite the objections of Einstein, Schrödinger, and de Broglie, it culminated with the triumph of the Copenhagen view.

From the early 1930s through the mid-1990s, most physicists regarded the question of the meaning of quantum mechanics as settled. This long dark age was punctuated by the important works of Bohm, Bell, Everett, and a few others, but most of the community of physicists paid little attention to these works or to foundational questions in general. One can see this from the fact that the crucial papers by those authors had very few citations into the mid-1970s, when the experimental tests of Bell's restriction began to be done. Even now, it is not uncommon to find very accomplished physicists who believe, incorrectly, that Bell proved all hidden variable theories must be wrong.* Until very recently, there were virtually no academic positions in physics departments for physicists focused on quantum foundations. The tiny community of specialists in quantum foundations either earned their tenure for other work, as Bell did, or, like Bohm, found places in out-of-the-way corners of the academic world. A few made careers in philosophy or mathematics, others by teaching in small undergraduate colleges.

It was the promise of quantum computing that began, just before the turn of this century, to open doors to people who wanted to work on quantum foundations. The idea that quantum mechanics could be used to construct a new kind of computer was broached by Richard Feynman in a lecture[1] in 1981. That talk, and other

* Rather than just ruling out local hidden variable theories.

early anticipations of the idea, seemed to make little impression until David Deutsch, originally a specialist in quantum gravity who held a position at Oxford, proposed in 1989 an approach to quantum computation in the context of a paper on the foundations of mathematics and logic.[2] In his paper, Deutsch introduced the idea of a universal quantum computer, analogous to a Turing machine. A few years later Peter Shore, a computer scientist working for an IBM research laboratory, proved that a quantum computer could factor large numbers much faster than a regular computer. At that point people began to take notice, because one application of being able to factor large numbers is that many of the codes now in common use could be broken.

Research groups began to spring up around the world, and they quickly filled with brilliant young researchers, many of whom had a dual research strategy in which they would attack the problems in quantum foundations while contributing to the development of quantum computing. As a result, a new language for quantum physics was invented that was based on information theory, which is a basic tool of computer science. This new language, called *quantum information theory*, is a hybrid of computer science and quantum physics and is well adapted to the challenges of building quantum computers. This has led to a powerful set of tools and concepts that have proved invaluable at sharpening our understanding of quantum physics. Nonetheless, quantum information theory is a purely operational approach that is most comfortable describing nature in the context of experiments, in which systems are prepared and then measured. Nature outside the laboratory hardly makes an appearance, and when it does, it is often, not surprisingly, analogized to a quantum computer.

The current renaissance of the field of quantum foundations/ quantum information is almost all for the good, not least because much of the theoretical work is anchored in real experiments. The drive toward quantum computation has led to many spin-offs which illuminate the foundational questions, such as *quantum teleportation*. This is a technology by means of which the quantum state of an atom can be transferred to a distant atom without being measured. If not quite up to science fiction's transporters, this technology is here now and is already playing a role. For example, it is used to make a new kind of code, which is unbreakable.

These developments have also deepened our appreciation for how quantum theory is structured. For example, a new type of approach, initiated by Lucien Hardy, seeks the shortest and most elegant set of axioms from which the mathematical formalism of quantum mechanics may be derived. Of these axioms, there are several that are unremarkable and tell us things that are true for every theory; then there is one into which all the strangeness of the quantum world is packed.

At the same time, there is little room, in a climate dominated by operational approaches, for old-fashioned realists in search of a completion of quantum theory that will explain what happens in individual events. Some of those realists are many-worlders, but there persists a small community of Bohmians. A handful develop theories of wave-function collapse. Those who try to push the search for reality beyond these established approaches are even fewer. Most of us in this class were originally specialists in other fields, some at the highest level of accomplishment, such as Stephen Adler and Gerard 't Hooft. We fit imperfectly into the lively field quantum foundations has become, especially as our concerns and ambitions—and the theories we develop to realize them—cannot be expressed in the operational language whose mastery is the sign of a professional quantum informationist. Still, we persist in our search for a realist and complete picture of the quantum world.

I believe that, as expressed by Lucien Hardy in the quote that opens this chapter, many physicists would prefer realism to operationalism, and would take an interest in the discovery of a realist approach to quantum theory that overcame the weaknesses of the existing approaches. If, during the present period, operational approaches dominate, this is partly due to the lack of a realist alternative which has the ring of truth.

The rest of this book is about the future of realist approaches to quantum physics. But before we dismiss the non-realist approaches, let's see if there is anything to be learned from the recent focus on them.

One lesson I've learned is that there are many different ways to express how the quantum world differs from the classical world of Newtonian physics. If you are happy taking an anti-realist point of view, there are a range of options. You can adopt Bohr's radical denial that science is anything other than an extension of common language we use to tell each other about the results of experiments we do. You can embrace an approach called *quantum Bayesianism*, according to which the wave function is no more than a symbolic representation of our beliefs, and prediction is a fancy word for betting. Another option is to embrace a purely operational perspective, which allows one to speak only of processes delineated by and sandwiched between preparations and measurements.

In all these the measurement problem is sidestepped, or rather, defined out of existence, because you cannot even pose the possibility that the quantum state describes the observers and their measuring instruments.

Several of the new proposals have at their core the concept that the world is made of information. This can be summarized in John Wheeler's slogan "it from bit," modernized as "it from qubit," where a *qubit* is a minimal unit of quantum information, i.e., a quantum binary choice, as in our story about pet preference. In practical terms, this program imagines that all physical quantities are reducible to a finite number of quantum yes/no questions, and also that evolution in time under Rule 1 can be understood as processing this quantum information as a quantum computer would. This means that the change in time can be expressed as the action of a sequence of logical operations applied to one or two qubits at a time.

John Wheeler put it like this:

> It from bit symbolizes the idea that every item of the physical world has at bottom—at a very deep bottom, in most instances—an immaterial source and explanation; that what we call reality arises in the last analysis from the posing of yes-no questions and the registering of equipment-evoked responses; in short, that all things physical are information-theoretic in origin and this is a participatory universe.[3]

The first time you hear this kind of view expressed, you may not be sure the speaker means it. But he does. Here is another, briefer quote: "Physics gives rise to observer-participancy; observer-participancy gives rise to information; information gives rise to physics."[4]

When Wheeler speaks of a participatory universe, he means that the universe is brought into existence by our observing or perceiving it. Yes, you might reply, but before we can perceive or observe anything we have to be brought into existence within and by the universe. Yes, says John. Both. Is there a problem?

Does this yield any insight? Some systems with a finite number of possible outcomes can be represented this way, and doing so does illuminate the physics: for example, the importance of entanglement in quantum physics can be brought into the foreground.

But other systems which have an infinite number of physical variables, such as the electromagnetic field, do not fit as easily within this program. Nonetheless, this quantum information approach to quantum foundations has had a good influence on diverse fields of physics, from hard-core solid-state physics to speculations on string theory and quantum black holes.

However, we should be careful to distinguish several different ideas about the relationship between physics and information, some of which are useful but also trivially true; others of which are radical and would need, in my view, more justification than they've been given.

Let's start by defining information. One useful definition was given by Claude Shannon, who may be considered the founder of information theory. His definition was set in the framework of communication, and contemplates a channel which carries a message from a sender to a receiver. These, it is assumed, share a language, by means of which they give meaning to a sequence of symbols. The amount of information in the message is defined to be the number of answers to a set of yes/no questions that the receiver learns from the sender by understanding what the message says.

Put this way, few physical systems are, or can be construed as, channels of information between senders and receivers who share a language. The universe as a whole is not such a channel of information. What is powerful about Shannon's idea is that a measure of how much information is transmitted can be separated from the semantic content, i.e., from what the message means. The sender and receiver share a semantics that gives meaning to the message, but you don't have to share that knowledge to measure the quantity of information carried. Still, without the shared semantics the message would not carry information. One way to see this is that to measure how much information a message carries, you need some information about the language, such as the relative frequencies with which different letters, words, or phrases occur in the linguistic community of those who speak that language. This information about context is not going to be coded into every message. If you don't specify the language, the Shannon information is not defined. This means, in particular, that the message has to be in a language that the sender and receiver share. A pattern of irregular symbols carries no information. So, to the extent that Shannon's measure of information depends on the language and other aspects of the context which are shared by the sender and receiver and not coded into the message, it is not purely a physical quantity.

One of the stubborn problems in the philosophy of language is to understand how speakers have intentions and convey meaning. That this is a hard problem does not mean that intentions and meanings are not part of the world. But they are aspects of the world that are dependent for their existence on the existence of minds. Shannon information is a measure of what goes on in this world of meanings and intentions. It is well defined even if we don't have a good understanding of how meaning and intention fit into the natural world, but it is nonetheless a part of that world.

Let me give an example to make this distinction clear. I hear drops of water falling intermittently from a leaky drainpipe after a summer rain. The pattern of the drips seems irregular, but it carries no message for me or anyone else. There is no sender, and I am no receiver; hence no information, in Shannon's sense, is contained in the drips. On the other hand, someone could use Morse code to send me a message via a sequence of short and long pauses between drips. The patterns between the two cases would differ in a way that reflects the presence or absence of an intention to convey meaning. The intent matters: information in this sense requires beings with the intention of conveying meaning. For a realist, who wants to know what the world is beyond what people know or understand, this is not a useful idea to apply to the atomic world.*

A less precise definition of information was given by Gregory Bateson, an English anthropologist, who called it "a difference that makes a difference." This idea is sometimes expressed instead as "a distinction which makes a difference." This is directly applicable to physics, where we might translate it as "If different values of a physical observable lead to measurably different futures of a physical system, that observable can be considered to constitute information." Put this way, almost every physical observable potentially conveys information. This definition would imply that "information" is present every time the values of two physical variables are correlated. But there is nothing profound about this, unless it is the appreciation of the interdependence of the different components of the physical world. And we already have measures of correlation. We can rename these "information," but a change of names that weakens the specificity of an idea is more likely to result in confusion than it is to bring about revolution in our conception of the world.

Computers process information in Shannon's sense. They take an input signal from a sender and apply to it an algorithm, which

* At this point I have to make an aside, which the non-expert reader may skip.

An expert might object to my characterization of Shannon information by pointing out that that quantity is equal to the negative of the entropy of the message. Entropy, they would argue, is an objective, physical property of nature, which is governed (when the system is in thermodynamic equilibrium) by the laws of thermodynamics. Hence, by virtue of its connection with entropy, Shannon information must be objective and physical.

I would answer by making three points. First, it is changes in the thermodynamic entropy, not the entropy itself, that come into the laws of thermodynamics. Second, as Karl Popper pointed out years ago, the statistical definition of entropy which Shannon information is related to is not a completely objective quantity. It depends on a choice of coarse-graining, which provides us with an approximate description of the system. The entropy of the exact description, given in terms of the exact state, is always zero. The need to specify this approximation brings an element of subjectivity to the definition of entropy. This is seen in the dependence of the entropy of a quantum system on a splitting into two subsystems. Finally, the attribution of entropy to a message is a definition, which defines entropy in terms of Shannon information.

transforms it into an output signal to be read by a receiver. Such contexts are very special. The choice of an algorithm to be embodied is a crucial part of the definition of a computation. Most physical systems are not computers, and the process by which the initial data at one time evolve to the data at a later time cannot always be explained in terms of the application of an algorithm or a sequence of logical operations.

Some authors appear to confuse and conflate the two definitions of information, which tempts them into wanting to describe nature as a computer and the relation between states of the world at different times as a computation. I am not convinced that such a radical hypothesis is justified.

This is not to say that some physical systems cannot be modeled to some degree of approximation by a computation, which is again trivially true. You can define approximations to the main equations of physics, such as those of general relativity or quantum mechanics, which can be coded as algorithms, which are then run on a digital computer. This is often a very useful way to get approximate solutions to the equations. But there is always an approximation involved.

The sound a symphony orchestra makes can be captured by a digitization, to an approximate degree, but this always involves an approximation, which truncates the range of frequencies. The full experience of listening to the orchestra live is never fully conveyed, which is why there is still an audience for performance as well as a market for vinyl, purely analog recordings. It is the same for physics: a digitization of Einstein's equations can be very useful, but it never captures all that the equations do.

Even if physics is not in general comprehensible as information processing, it may be asserted that the quantum state represents not the physical system, but the information we have about the system. Rule 2 certainly makes it seem to be the case, because the wave function changes abruptly just when we gain new information about the system. But if the wave function represents the information we have about a system, then the probabilities quantum mechanics predicts must be seen as subjective, betting probabilities. This viewpoint can be developed by understanding Rule 2 as an update rule by which our subjective probabilities for future experiments change as a measurement is made. This is what is called quantum Bayesianism.[5]

A RATHER ELEGANT APPROACH, which also sees the quantum state as conveying information that one system has about another, is called *relational quantum theory*. According to this view, which sits between operationalism and a form of realism, quantum states are associated with splits of the universe into two parts, observer and observed, and represent what the former can know of the latter. This idea had its roots in quantum gravity, and arose out of conversations between Louis Crane, Carlo Rovelli, and me in the early 1990s.

Our inspiration was a very elegant body of mathematical descriptions of very simplified cosmologies, which Crane and other mathematicians had developed, called *topological field theories*. In these theories there is no quantum description of a whole universe. There is no quantum state describing the universe as a whole. Instead, there is a quantum state for each way of dividing the universe into two subsystems. These can be thought of as carrying the information that an observer on one side of the divide could have about the quantum system on the other side.

This reminded us of Bohr's insistence that quantum mechanics requires a split of the world into two parts, one classical, the other quantum, and that any split will do. The models Crane and other mathematicians had studied took Bohr's philosophy a step further, for there were two quantum states for every boundary—one for each side. This is because there are two ways to read each split. If Alice lives on one side and Bob lives on the other, then Alice will see herself as a classical observer, measuring a quantum Bob, but Bob will see things the other way around.

The models were very simple, so that there was only one question that could be asked, which was: How similar were the two views? What is the probability that Alice's quantum description of Bob will be the same as Bob's quantum description of Alice? The mathematicians set up their theories so that the answer was the same however the universe was split. In that case, the probability of one side's view resembling the other side's view measures something universal, which would characterize how that universe is connected, i.e., what mathematicians call the universe's topology. This is why they were called topological field theories.

Crane brought these model universes to Rovelli's and my attention because he saw that the mathematical structures involved could be extended to encompass loop quantum gravity. He turned out to be right about that, but that is another story. Crane also proposed that the new mathematics offered a way to extend quantum mechanics to the universe as a whole. He was right about that too, and the result was relational quantum theory.

We were each inspired to apply this idea to quantum theory in general, and we each published a version of it.[6] Rovelli's formulation was the most general, and has become the best known, so I'll describe his formulation of the idea.

Bohr taught that quantum physicists must speak always of two worlds. We observers live in the classical world, but the atoms we study live in a quantum world. The two worlds satisfy different rules. In particular, objects in the quantum world can exist in superpositions, but observable properties of things in the classical world always take sharp values, and so cannot be superposed. Bohr's point is that both worlds are necessary for science.

The instruments we use to manipulate and measure the atoms live at, and in a sense define, a boundary between them and us. Bohr emphasized that the placement of this boundary is arbitrary, and could be drawn differently for different purposes, so long as it divides the world into two domains.

Let us think of the Schrödinger's cat experiment. One way to draw the boundary is to consider the atom and photon as the quantum system, keeping the Geiger counter and cat in the classical world. In this picture the atom may exist in a superposition, but the Geiger counter will always show a definite state—either YES, it saw a photon, or NO, it did not. But we can redraw the boundary, including the detector in the quantum world. In this picture, the cat is always either dead or alive, but the Geiger counter may be in an entangled superposition with the atom. Or, and this was Schrödinger's point, you can instead draw the boundary to coincide with the walls of the box, so the cat is now also part of the quantum system and can exist in entangled superpositions with the atom and Geiger counter. The classical world then includes a friend of ours, Sarah, who opens the box and looks in. Sarah, we presume, is macroscopic and classical and so can be treated as always being in a definite state. From her viewpoint, Sarah experiences herself to be on the classical side of the boundary, so, according to her, she always sees the cat to be either alive or dead.

Eugene Wigner suggested we take this fable one step further

and consider that the quantum system includes also our friend Sarah, together with the box, the cat, and the box's other contents.* I remain outside the boundary, so I see Sarah become part of a superposition of entangled states. In one part of the superposition the cat is alive and Sarah sees it to be alive, while in another part of the superposition the cat is dead and she sees it to be dead.

Thus we have five different ways to divide the world into quantum and classical, where by quantum we mean it could be in a superposition, while classical means that physical quantities always have definite values. These different descriptions appear to disagree with each other. We see Sarah to be in a superposition whereas she always sees herself to be in a definite state.

Rovelli's proposal is that these are all equally correct, partial descriptions of the world. All are part of the truth. Each gives a valid description of a part of the world, defined by a boundary. Is Sarah truly in a superposition, or does she definitely see and hear a live cat? Rovelli would like to not have to choose between these. He insists that a description of physical events and processes is always made with respect to some particular way of drawing the boundary between quantum and classical. He posits that all ways of drawing the boundary are equally valid and all are part of the total description.

Simply put, Rovelli would say that it is true, *from Sarah's point of view,* that the cat is alive, and it is also true, *from my point of view,* that Sarah is entangled in a superposition of "seeing dead cat" and "seeing live cat."

Is there any truth that is not qualified by a point of view? My understanding is that Rovelli would say no. In the story as I've told it, Sarah and I agree that she opened the box and inspected the cat,

* This step of the argument is called "Wigner's friend."

even if we don't agree on the outcome. But it could have been the case that Sarah's decision to open the box depended on the outcome of a quantum event such as the decay of an unstable atom, in which case I may describe Sarah as being in a superposition of having looked in the box and not. But Sarah will experience one or the other.

Notice that there is a weak kind of consistency, in that my description of Sarah does not preclude hers. Notice also—and this is central—that every way of drawing a boundary splits the world into two incomplete parts. There is no view of the universe as a whole, as if from outside of it. There is no quantum state of the universe as a whole.

If relational quantum theory had a slogan, it would be "Many partial viewpoints define a single universe."

This proposal can be seen through various lenses. A pragmatic operationalist sees each way to divide the world in two with a boundary as defining a system that can be treated with quantum mechanics. Each choice results in a description, which contains all the information that an observer on the classical part of the boundary can have about the quantum system on the other side of the boundary. For such an operationalist, the collection of quantum states contains the information that an observer can have at each level, defined by a boundary that sets her apart. Each observer uses a quantum state to code the information they have about the system on the other side of their boundary; these different states are different because they are descriptions of different subsystems.

Seen through this operational lens, relational quantum mechanics has something in common with Everett's original relative state interpretation. Each describes the world in terms of contingent statements that code correlations between different subsystems, which are established when they interact.

But this is not the way that Rovelli sees relational quantum mechanics. Rovelli wants to call his view realism, but it means something different from naive realism, as I have used the term so far. For him, reality consists of the sequence of events by means of which a system on one side of a boundary may gain information about the part of the world on the other side. Thus, we can say that Rovelli is a realist about causation. This reality is dependent on a choice of boundary, because what is a definite event—something that definitely happened for one observer—could be part of a superposition for another. Thus, Rovelli's realism is different from naive realism, according to which what is real consists of events that all observers will universally agree took place.

Rovelli denies that that kind of naive realism is possible in our quantum world, so he proposes we adopt his radically different version of realism, according to which what is real is always defined relative to a split of the world that defines an observer. Rovelli uses very different words than Bohr, and achieves a formulation which is more precise, but the two employ a similar logic, which denies the possibility of naive realism about quantum systems.

ANOTHER APPROACH WHICH DENIES that naive realism is possible is based on elevating the category of the possible—things that might be true—to the world of the real. Naively, when we say that something is possible, such as that my son's lizard might become pregnant in the next year, we mean it is among the things that might happen. When something possible happens it becomes part of the real; but till then it is not real.

Language and logic reflect the very different status of the possible, and distinguish it from the real. The law of the excluded middle says that something real cannot simultaneously have a property and not have it. Our neighbor's bunny rabbit cannot be both gray and not gray. But possible states of affairs have no such constraint. The rabbit our friend will buy at the pet store next week might possibly be black and it also might possibly be white.

In real life the actual and the possible have an asymmetric relation. The real existence of our neighbor's daughter makes a rabbit a possible future pet for their family. So what is possible is influenced by what is real. But knowledge of the possible, while helpful, is not strictly speaking necessary for working out what will be real; to the extent that the laws of Newtonian physics are deterministic, all you need to predict the actual future is a complete description of the actual present.

SEVERAL WRITERS, beginning with Heisenberg and including my teacher Abner Shimony, have proposed that the world of the possible has to be included as part of reality—because in quantum physics the possible influences the future of the actual. This view has been recently developed by my friend Stuart Kauffman, in collaboration with Ruth Kastner and Michael Epperson.[7]

There is no way to describe this view that doesn't cause some tension with ordinary language usage, but keep an open mind and I'll aim to be clear. We start by stating that there are two ways for a circumstance to be real. It can be actual, which means that it is part of the world in the same way that a Newtonian particle has a definite position. But something real can also be "possible" or "potential"; this is the status we assign to properties that are superposed in the wave function, such as a leftist having equal cat and dog preference, or a particle which could go through the left slit or the right slit, or Schrödinger's cat being both alive and dead.

Things that are real but possible don't obey the law of the

excluded middle, but they are to be considered part of the real because they can influence the actual. This is, according to this perspective, what is different and new about quantum physics. According to Kauffman and his coauthors, experiments are processes that convert potentialities to actualities. Thus, Schrödinger's cat is potentially alive and potentially dead, not in the sense of something that is one or the other, but about which we are ignorant, and not in the sense of some undetermined state of affairs, but because its actual reality consists of this potentiality for one or another to be realized by an experiment.

The fact that experiment plays a distinct role in converting the possible to the actual, with probabilities given by the Born rule, is enough to tell us that this is not a naively realist perspective, i.e., a description of the world as it would be in our absence, in which experiment cannot play any role. But it is a direction, perhaps, to be developed, if realism fails.

Here is a way we might develop the view that the possible is part of the real. Bring in time, and let us take the view that the present moment and the flow or passage of moments are real and fundamental.* Part of what I mean by this is that there is an objective distinction between the past, present, and future. In such a view, the present is real. The present consists of events which have happened, but which have yet to give rise to the future events that will be their replacements.

The past consists of those events which were once present and real. They no longer exist, although their properties can be captured and remembered in presently existing structures.

The future is not real. Moreover, the future is slightly open, in the sense that rare novel events with novel properties may happen

* As I argue in *Time Reborn*, and in *The Singular Universe and the Reality of Time*, with Roberto Mangabeira Unger.

every once in a while. (See my *principle of precedence* below.) But if for a moment we ignore that possibility, then there does exist in the present a finite set of possible next steps, which are possible next events and their properties.

Given the present state of the world, not everything can happen in the next time step. Those events that might be next Kauffman calls the *adjacent possible.* The possible near-future events that make up the adjacent possible are not yet real, but they define and constrain what might be real.

The adjacent possible of Schrödinger's cat includes a live cat and a dead cat. It does not include a brontosaurus or an alien dog. So the elements of the adjacent possible have properties, even if the law of the excluded middle does not apply to them. As objects with properties, there are facts of the matter about them. This is the sense in which we may say that a small part of the possible may be considered real.

This starts to make sense. Not everything that is possible is real. But a small part of the possible has definite properties that justify assigning it to a new category of the real and possible.

THERE ARE ALSO RECENT DEVELOPMENTS on the magical realism side. Back in the 1990s Julian Barbour proposed a quantum theory of cosmology that has many moments rather than many worlds.[8] This has been revived in a recent proposal by Henrique Gomes. As we are not concerned with technical details, I'll describe the original approach of Barbour, but most of what I'll have to say applies to Gomes's version[9], as well as more recent work of Barbour and his collaborators.

A moment, for them, is a configuration of the universe as a whole. These configurations, according to Barbour and Gomes, are relational configurations, which code all the relations that can be captured in a moment, such as relative distances and relative sizes.

We seem to experience time passing as a smooth flow of moments. Barbour insists that the passage of time is an illusion and that reality consists of nothing but a vast pile of moments, each a configuration of the whole universe. You now are experiencing a moment. Now you are experiencing a different moment. According to Barbour, both moments exist eternally and timelessly, in the pile of moments. Reality is nothing but this frozen collection of moments outside time. Each experience of a moment also exists timelessly—as part of its moment. The fleeting aspect of a moment is in reality just an aspect of the moment, a feature it has eternally.

The moments all coexist, and each is a configuration of the whole universe. But there is an important way they can differ. The pile can have more than one copy of a configuration, and the number of copies may vary from many copies to none at all.

Barbour hypothesizes that we are equally likely to be experiencing any of the moments in the pile. But since some are more common than others, there is structure to our experience, as we are most likely to experience the more common moments.

The collection of moments is structured so that the most common moments are those configurations that, to some degree of approximation, can be strung together as if they were a history of the universe generated by a law. This gives us the illusion that laws are acting, but there are no laws generating histories, and indeed no history. Reality is just the vast collection of moments.

Barbour hypothesizes that the most common moments contain structures which speak to us of other moments. A book, even while frozen forever in a moment, may tell stories that are only comprehensible as a sequence of events that played out over time. A book has a publication date, which references a happy event (at least for its author) sometime in the past. And it was brought into existence by a printing company, a publishing company, and a paper mill, each of which has a history, which evokes more stories.

Barbour calls objects like books, which contain eternally frozen, momentary structures that are pointers to other moments, *time capsules*. Anything that is, or contains, a record, such as a DVD or a video file, is a time capsule. So it can be any built structure or manufactured object. Indeed, it can be any living thing.

For most of us, the fact that the natural world is chock-full of time capsules is evidence that time is real and fundamental. Events are ordered in time because past events cause present events. But according to Barbour, even the impression we have of living within a flow of moments is an illusion. All the memories, records, and relics we have that give the impression that there was a past are, in fact, aspects of a present moment. Each moment lives eternally in the pile of moments.

An unordered pile of moments, which is all that makes up a Barbourian universe, might easily contain few moments with time capsules. Why then is almost every moment of our universe full of them?

To elucidate our world, Barbour has to explain what determines which configurations are common, having many copies in the pile, and which are less common, or altogether absent. This is dictated by an equation, which is the only law that acts to structure the pile. It does so by choosing which configurations are represented in the pile, and by how many copies. This is a version of Schrödinger's equation, but one with no explicit reference to time. It is called the Wheeler-DeWitt equation; we can call it Rule 0. This equation chooses as solutions piles of moments which are populated by those that can be strung together to permit the illusion of history to emerge.

If this is right, then the passage of time is an illusion, which is due to a present moment containing the experience of memories of the past. Causality is also an illusion.

These "many moments" theories are realist, in that they take a stand on what is real, which is the timeless collection of moments. But these theories are beyond naive realism in that they posit a real world enormously different from the time-bound world we experience, in which we perceive a succession of moments, one at a time.

The lesson I draw from these theories is that to extend quantum mechanics to a theory of the whole universe, we have to choose between space and time. Only one can be fundamental. If we insist on being realists about space—as Barbour and Gomes do—then time and causation are illusions, emergent only at the level of a coarse approximation to the true timeless description. Or we can choose to be realists about time and causation. Then, like Rovelli, we have to believe that space is an illusion.

There is much more that could be said about these recent nonrealist and magical-realist perspectives. But the bottom line is that if your interest is pragmatic, and you want to use quantum theory to understand questions other than those arising from quantum foundations, any of these will serve to frame your calculations and the explanations you draw from them. But if you want to solve the measurement problem in a way that gives a detailed description of what goes on in an individual physical process, nothing but a realist description will do.

THIRTEEN

Lessons

The main message of this book is that however weird the quantum world may be, it need not threaten anyone's belief in commonsense realism. It is possible to be a realist while living in the quantum universe.

However, simply affirming realism is not enough. A realist wants to know the true explanation for how the world works. There would be no sense in believing that the world has a detailed explanation, and not being interested in what that explanation is. Thus the next question to ask is whether any of the available realist versions of quantum physics are compelling as true explanations of the world. That is, are we done, or do we have more work ahead? Unfortunately, I believe the answer is that, so far, none of the well-developed options are convincing. All the realist approaches that have so far been studied have serious drawbacks. To explain why, let me review the available options, with a focus on the strengths and weaknesses of each.

PILOT WAVE THEORY

Pilot wave theory completes quantum mechanics by providing additional degrees of freedom which, together with the wave function, fully specify what is going on in an individual physical system. These are the particle trajectories. We called these hidden variables, but that is perhaps not the best way to talk, as the particles are, after all, what is observed. A better way to describe the options is to use the term "beables," as suggested by John Bell. Realists want a theory to take a stand about what really exists; these are the beables. In pilot wave theory the waves and particles are both beables.

Pilot wave theory solves the measurement problem, because the particle always exists and it is always somewhere. When an experimental device looks for the particle, it finds it where it is.

The equations of pilot wave theory are deterministic and reversible, which argues for the completeness of the theory. Probability is explained by our ignorance of the initial positions of the particles, just like in other applications of probability to physics. The Born rule, the relationship between probability and the square of the wave function, is explained by the demonstration that this is the only stable probability distribution, and all others evolve to it.

In addition, pilot wave theory is complete and unambiguous. Some of the other modifications of quantum mechanics come with new free parameters, which may be adjusted to hide various embarrassments and protect the theory from experimental disproof. Pilot wave theory has no additional parameters and allows no choices to be made. This is a very important point in its favor.

Because it gives a clean, unambiguous, and explicit description of the quantum beables, pilot wave theory continues to be a popular option within the small community of quantum realists. Partly this is because there remains a lot to do to develop the applications of the theory. It is one thing to demonstrate generally that the predictions of pilot wave theory and conventional quantum mechanics will often agree, but it is another to see how this works out in detail. Physicists like to have well-defined problems to work out, and pilot wave theory offers no shortage of these.

There are challenges for pilot wave theory. If it is to replace quantum mechanics, it must do so in all the contexts in which the usual theory works. This includes relativistic quantum field theory, which is the basis of the standard model of particle physics. There has been good work done on this, but important questions remain unresolved. There have also been very interesting explorations of pilot wave theory applied to quantum gravity and cosmology, but these are far from definitive.

But the most important aim of research in pilot wave theory must be to discover and open up domains where experiments will distinguish the new theory from the older one. Here there is exciting work being done on the cosmological scale by Antony Valentini and others.

At the same time, there are several reasons pilot wave theory is not entirely convincing as a true theory of nature. One is the empty ghost branches, which are parts of the wave function which have flowed far (in the configuration space) from where the particle is and so likely will never again play a role in guiding the particle. These proliferate as a consequence of Rule 1, but play no role in explaining anything we've actually observed in nature. Because the wave function never collapses, we are stuck with a world full of ghost branches. There is one distinguished branch, which is the one guiding the particle, which we may call the occupied branch. Nonetheless, the unoccupied ghost branches are also real. The wave function of which they are branches is a beable.

The ghost branches of pilot wave theory are the same as the branches in the Many Worlds Interpretation. In both cases they are a consequence of having only Rule 1. Unlike the Many Worlds Interpretation, pilot wave theory requires no exotic ontology in terms of many universes, or a splitting of observers, because there is always a single occupied branch where the particle resides. So there is no problem of principle, nor is there a problem of defining what we mean by probabilities. But if one finds it inelegant to have every possible history of the world represented as an actuality, that sin is common to Many Worlds and pilot wave theory.

A perceptive reader might be troubled by this similarity to the Many Worlds Interpretation. Assuming that its proponents do succeed in giving Everettian quantum mechanics a sensible physical interpretation via decoherence, subjective probabilities, and the works, couldn't we apply exactly the same interpretation to the branching wave function of pilot wave theory—and simply ignore the particles? The answer is yes, you can ignore the particles, and then you are squarely back in Everett's multiverse. This brings up a hidden, perhaps unconscious assumption made by the adherents of pilot wave theory, which is that the reality that we observers perceive and measure is composed of matter constructed from the particles of pilot wave theory.

Just because both the particles and the waves are beables in pilot wave theory does not make them equivalent. To make sense of pilot wave theory we must privilege the particles and postulate that the world we perceive consists of the particles. The waves are there in the background, but their role is to guide the particles—they are not perceived directly, and only affect our observations through their role as guides.

From the point of view of either explaining or predicting the world, the ghost branches play little role in pilot wave theory. The probability that a ghost branch of a macroscopic system will interfere with the occupied branch, changing the future of that system, is truly tiny. It is tempting then to introduce some mechanism to prune back the ghost branches. This would be a combination of pilot wave theory and spontaneous collapse models. I am not aware of any work in this direction, but it seems an interesting avenue to explore.

This brings up another issue with pilot wave theory, which is that there is an asymmetry of causes. The wave function guides the particle, but the particle has no influence back on the wave function. This is unlike the way causes work in ordinary physics. In nature, and so in most of physics, causes are usually reciprocal. Everything you push on pushes back. This is due to Newton's third law, which states that every action is met by an equal and opposite reaction. It is then very strange that the particle cannot influence the wave function. The lack of a reciprocal effect strongly suggests something is missing.

Even if the ghost branches can often be ignored, they can't always be. Some clever experiments have been devised which show that the branch of the wave function which the particle doesn't take can influence the future as much as the occupied branches.[1] These tricky cases involve two quantum particles which interact with each other, such as an atom and a photon.

According to pilot wave theory, the atom is both a particle and a wave. Let's call them the atom's particle and the atom's wave. The photon is also a particle and a wave, and, likewise, we'll call them the photon's particle and the photon's wave. In each case, the wave guides the respective particle. But suppose we set things up so that the photon is to collide with the atom. Which entity interacts with which?

You might be tempted to suppose that the atom's particle

collides with the photon's particle. But that turns out to be wrong. The two particles are each invisible to the other. They will easily pass right through each other. Instead, what happens is that the two waves interact and scatter off each other. Then, as the waves retreat from their collision, the atom's wave pulls the atom's particle with it, while the photon's wave likewise pulls the photon's particle away.

But whether a wave function scatters another wave function doesn't depend at all on whether it is an occupied branch or a ghost branch. This has some pretty weird consequences, but they are equally weird for conventional quantum mechanics and pilot wave theory. For example, it can appear that a particle bounces off the empty ghost branch of another particle's wave function.

The fact that wave functions bounce off wave functions doesn't count against pilot wave theory. Indeed, it shows that the theory works even in such counterintuitive situations; this should strengthen our confidence in it. But it teaches us the cost, which is to give up comfortable pictures in which the particles are the main story and the ghost branches are discounted.

The fact that the particles are guided by the wave functions has other weird consequences, one of which is that the motions the particles make in response to their guidance by the wave function fail to conserve momentum and energy. The particles behave like UFOs in bad science fiction movies—for example, they can sit still for hours, which is what they do in states of definite energy, and suddenly jump up and run away in response to changes in the guiding wave function.

This did not shock de Broglie, and it doesn't perturb his modern followers, such as Valentini. They understand it has to be that way, because part of the guidance equation's job is to bend the paths of particles around obstacles and through slits, to reproduce the diffraction of light, and a particle that alters its direction without colliding with another particle is one that changes its momentum. But this was a deal breaker for Einstein and, I would guess, it has been for others.

If one averages a system that is in quantum equilibrium over many possible trajectories of the particles, then on average momentum and energy are conserved. This is one reason I've come to favor formulations in which the probabilities refer to ensembles of particles that really exist. I will be discussing these in the next chapter.

Pilot wave theory offers a beautiful picture in which particles move through space, gently guided by a wave, which also flows in space. The reality is a bit less intuitive. When applied to a system of several particles, the wave function doesn't flow through space; it flows on the configuration space, which is multidimensional and thus hard to visualize. And, as I've emphasized, the particles are not your grandmother's little round spheres—they react to things near and far, including sudden nonlocal influences transmitted through the guidance equation. Still, the particles can do nothing else if pilot wave theory is to reproduce the results of quantum mechanics.

A third problem with pilot wave theory is that there is a strong tension with relativity theory. This is because of nonlocality. The experimental tests of Bell's restriction tell us that any attempt to go beyond the quantum, to give a description of individual events and processes, must explicitly incorporate nonlocality.

This nonlocality must somehow be coded into the pilot wave theory, because that theory is a completion of quantum mechanics and agrees with its predictions. And indeed, nonlocality is built in. How can that be? Let us consider a system of two entangled particles, which are very distant from each other. The secret is that the quantum force that one particle experiences depends on the position of the other particle. This dependence remains even if the two particles are very far from each other.

As a result, if one could follow the trajectories of the individual quantum particles, one could see that entangled particles are influencing each other nonlocally (i.e., at a distance). Because we normally measure only average positions and average motions, this incessant nonlocal influence is washed out by the randomness of the quantum motions. But it is there explicitly in the way the wave function guides the particles, and one can contemplate experiments which might be able to observe it.

The alert reader may be hearing alarm bells going off. This nonlocal communication of forces over a distance requires us to objectively speak of events that are distant from each other, but are nonetheless simultaneous. Such an instantaneous effect at a distance directly contradicts special relativity, which tells us that there is no absolute notion of simultaneity for distant events. This is indeed a problem, and as a result there is a tension between special relativity and pilot wave theory.

In particular, the guidance equation, which is the source of the nonlocal forces, is inconsistent with relativity. It requires for its definition a preferred frame of reference, which defines an absolute notion of simultaneity. In practice, the conflict is blunted because the randomness of quantum physics implies that, so long as one stays in quantum equilibrium,* one cannot directly observe the nonlocal correlations in an experiment. Nor can we send information faster than light. If we don't look too closely at what is happening in individual systems, pilot wave theory maintains an uneasy coexistence with relativity. But then again, the whole point of pilot wave theory is that it enables us to look more closely.

* Which I defined back in chapter 8.

At the present time there is work in progress aimed at extending pilot wave theory to relativistic field theory, so we cannot give a definitive picture as to how this tension between relativity and pilot wave theory resolves.[2]

WAVE-FUNCTION COLLAPSE

The spontaneous collapse hypothesis also serves us well as a realist description of the quantum world in terms of beables. According to this picture, there are no particles—only waves—but those waves occasionally interrupt their smooth flow to suddenly collapse into particle-like concentrations. From there, the wave flows and spreads out again. Because the wave has this peculiar behavior, it mimics particles when needed, and thus is the only beable.

The collapse models also solve the measurement problem, because the collapse of the wave function is posited to be a real phenomenon. For atomic systems this is rare. But the rate of collapse grows rapidly with the size and complexity of the system, so there is no chance for superpositions and entanglements to survive for macroscopic systems. Superpositions and entanglements are destroyed by the collapses, and so are limited to the atomic domain. This solves the measurement problem, because the wave functions of the measuring instruments are always collapsed somewhere definite. It also gets rid of the ghost branches.

The pilot wave theory and spontaneous collapse models are not just two different interpretations of quantum mechanics. They are distinct theories, which each make some predictions that differ from those of quantum mechanics. Yet when it comes to the behavior of atoms and molecules, they agree with each other, and with conventional quantum mechanics, to much better precision than the experiments can detect. So, up until this point they cannot be distinguished experimentally from each other or from quantum mechanics. Pilot wave theory, however, predicts that superposition and entanglement are universal and should be in principle detectible in any system, no matter how large or complex. This is challenging to test experimentally, because one has to fight the tendency for a system of many particles to decohere, as the many interactions with the system's environment randomize the phases* of the wave function. In principle it can be done, and, indeed, experimentalists are continually expanding the domain of quantum phenomena.

But if the wave function undergoes spontaneous collapse, as soon as that happens the game is up. If spontaneous collapse is right, no experimentalist will ever be able to superpose two wave functions of a large, complex system.

Another difference between spontaneous collapse and pilot wave theory lies in their attitude toward time. The laws of pilot wave theory are reversible in time, just like the laws of Newtonian dynamics. Spontaneous collapse is irreversible, like the laws of thermodynamics.

The theories of wave-function collapse have some of the same drawbacks as pilot wave theory. In particular, the collapse is instantaneous, but takes place everywhere at once, creating a severe conflict with relativity. As with pilot wave theory, the precise law requires a preferred frame of reference to be specified and therefore contradicts relativity theory. And, as in that case, there is some work that indicates that the conflict can be managed, so that in the domain where the theory agrees with quantum mechanics, the violations of relativity theory are very small.

---

* The *phases* of a wave function refers to the locations of the peaks and troughs, and the patterns they form.

Another drawback of some collapse models is the fact, already mentioned, that energy is not conserved. Still another is that this defect can be minimized by tuning a free parameter. To my understanding, the ability to tune parameters to ensure agreement with an experiment is a weakness, as it suggests the theory is contrived to hide an essential tension in its construction.

Indeed, collapse models come in several versions, and there is some freedom to modify them and tune new parameters. That is why they are called models, while pilot wave theory, having no freedom to adjust anything, is a theory.

Among the various issues we have discussed, it is impressive that all the hidden variable theories which have been proposed conflict with special relativity. The reason is simple. If one wants a complete description of individual processes, that description must, because of the experimental tests of Bell's restriction, be nonlocal, and that requires a preferred simultaneity. Averaging over individual cases gives one probabilities, and since these agree with the probabilities predicted by quantum mechanics, there is no manifest contradiction with special relativity, because information cannot be sent faster than light. But for a realist the conflict is nonetheless present because reality is made of individual cases. We see this clearly in pilot wave theory and in spontaneous collapse models.

Nor can one escape this dilemma by giving up the ambition of going beyond quantum mechanics, for the conflict is present in quantum mechanics itself. When the wave function collapses following Rule 2, it does so everywhere at once.

No problem in physics has given me more pain, and kept me up more nights, than this conflict between commonsense realism applied to the atomic domain and the principles of special relativity.

To my mind, the most important reason to be skeptical about both pilot wave theory and collapse models is that they make little contact with the other big questions in physics, such as quantum gravity and unification.

At minimum, both approaches provide proof of concept that we can be realists about quantum physics. But neither has the ring of truth. There is more work to do to discover a realist completion of quantum mechanics that avoids the pitfalls of the existing theories while offering solutions to the other key questions in physics, and so gives us a platform on which to rebuild physics.

THERE HAVE BEEN SOME new proposals of realist quantum theories, none of which are, to my mind, completely convincing either. But they contain some intriguing ideas.

RETROCAUSALITY

A recent realist approach to quantum mechanics is *retrocausality*, which supposes that causal effects can go backward as well as forward in time. Usually the effect follows the cause, but, the proponents of this view argue, sometimes the effect precedes the cause. By zigzagging backward and forward in time, a chain of causations can appear nonlocal, as we see in figure 10 on the following page. The trick is easy. If we can go backward in time at light speed, and then forward, we can end up at an event simultaneous with, but far from, where we started. So in a theory with causation both in the future and in the past, we can aim to explain nonlocality and entanglement.

This kind of approach has been advocated by Yakir Aharonov[3] and colleagues. Another version, called the transactional interpretation, has been proposed by John Cramer and Ruth Kastner.[4] Huw

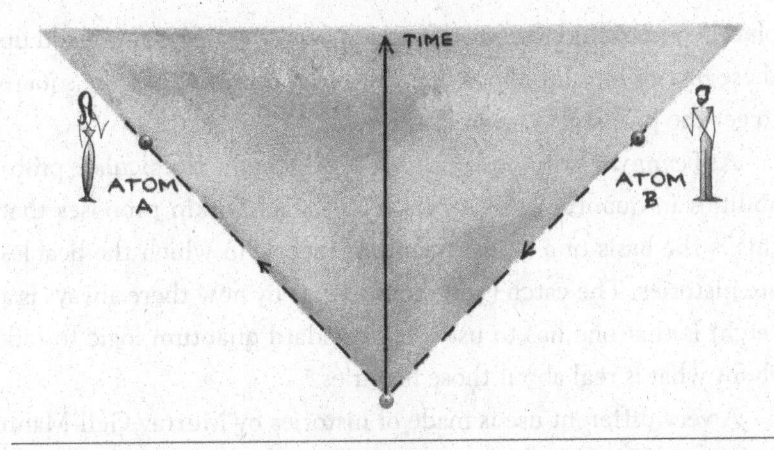

FIGURE 10. RETROCAUSALITY The two atoms travel to the future, one to the left and one to the right. But a causal influence can travel from the location marked atom B back to the point in the past from which the atoms originated, and then forward to the point at atom A. Thus the effect at atom A appears to be simultaneous with its cause at atom B.

Price has published an argument that any time-symmetric version of quantum mechanics must rely on retrocausality.[5]

APPROACHES BASED ON HISTORIES

An ancient idea holds that what is fundamentally real is not things, but processes; not states, but transitions. This bold idea underlies several approaches to quantum physics. They arise from a discovery Richard Feynman made while he was still a PhD student. Feynman formulated an alternative way of expressing quantum mechanics that eschews the description of nature in which quantum states change continuously in time. Instead, we compute the probability for the system to make a transition between an earlier configuration and a later configuration. We do this by considering all the possible histories that might have taken the system between the two configurations. The theory assigns to each history a quantum phase,* and to find the wave function for the transition, we add up these phases for all the possible histories. Then we take the square to get the probability, as in Born's rule.

As Feynman proposed it, this is just a scheme to calculate probabilities in quantum mechanics. But Rafael Sorkin proposes that this is the basis of a realist quantum theory, in which the beables are histories. The catch (you should know by now there always is a catch) is that one has to use a nonstandard quantum logic to talk about what is real about those histories.[6]

A very different use is made of histories by Murray Gell-Mann and James Hartle,[7] who maintain that the reality we experience is just one of many equally consistent and equally real histories. The idea is that if different histories decohere, they can't be superposed; thus they can be thought of as alternative histories. Gell-Mann and Hartle, along with Robert Griffiths and Roland Omnès, formulated this idea as the *consistent histories approach* to quantum mechanics.[8] A key result of this approach was that a history obeying Newton's laws of classical physics would be part of a family that would decohere. These decoherent histories could be treated as if they were alternative real histories. However, the converse was shown not to be the case by Fay Dowker and Adrian Kent, who demonstrated that there are many classes of histories that decohere which are not related to Newtonian physics.[9]

None of these history-based theories satisfy my desire to have a naively realist description of the world. I have nothing against a realism in which what is real is processes rather than states, happenables rather than beables. But in the approaches I've just mentioned, you end up computing not what happened, but only the probabilities for what happened. And the relationship between the histories

* That is, a complex number.

posited by the theory and the probabilities we observe are always related by Born's rule, which suggests that those histories represent possibilities and not actualities.

MANY INTERACTING CLASSICAL WORLDS

Here is another contemporary realist formulation of quantum physics.[10] Assume that our world is classical, but it is just one of a very large number of classical worlds, which exist simultaneously. These worlds are similar to each other, in that they have the same numbers and kinds of particles. But they differ as to the positions and trajectories of the particles.

All these worlds obey Newton's laws, with a single change, which is that, in addition to the usual forces between the particles in a single world, there is a new kind of force, which involves an interaction between the particles in the different worlds.

When you throw a ball, it responds to the force from your arm as well as the gravitational attraction of the Earth. At the same time, a large number of similar copies of you, each in their own world, throws a ball. Each of these balls has a slightly different starting point and trajectory. The different balls reach out to each other from their separate worlds and interact with each other. These new, inter-world forces are tiny, but the result is that each ball is jiggled a bit as it travels. You only observe the ball in your universe, so you can't account in detail for the jiggles. Thus there appears to be a random fluctuation which slightly disturbs the flight of your ball. The result is that you have to introduce a random, probabilistic element into any predictions you may make of your ball's motion. This probabilistic description is quantum mechanics.

This is called the *many interacting worlds theory*. To make it work out in detail, you have to choose the forces between the worlds very carefully. To get quantum mechanics to emerge, that force must be unlike any force we know about. It has to involve triplets of worlds, so there is a jiggle on your ball which depends on where two other balls are, each in their own worlds.

One great advantage of this formulation is that it's been extremely useful as a basis for detailed and highly accurate computer calculations of the chemistry of molecules.[11]

I am not going to suggest we take this as a serious proposal about nature. But it serves as another example of a realist version of quantum physics.

SUPERDETERMINISM

Not everyone working on quantum foundations accepts the conclusion of Bell's theorem that locality is violated in nature. There are several loopholes, most of which have been ruled out by experiment. One loophole which is not as easy to rule out is based on an idea called *superdeterminism*. Recall Aspect and colleagues' experiment disproving Bell locality, which we talked about in chapter 4. Two observers, distant from each other, each choose a direction along which the polarization of the photon on their side will be measured. The proof that locality is violated relies on an assumption that these two choices are made independently.

But, strictly speaking, the two events in which these choices are made are both in the causal future of some events in their past. We just have to go back in time far enough until we find events whose causal futures include both events when the choices of which polarization to measure were made. So we could include such events in the past, whose causal future includes the whole experiment, as necessary parts of the experiment. You could then imagine that the angles chosen on each side were both specified by someone setting up very carefully the initial conditions in the past of both. The philosophy of superdeterminism asserts that the universe evolves deterministically so that all such correlations were fixed long ago, in the big bang.

Several physicists have proposed that if we assume that the initial state of the universe was chosen extremely delicately (by whatever agency can be recognized as setting the initial conditions), all the entangled pairs that would ever be measured could be determined to be set up in such a way as to mimic the results that are usually taken as confirming nonlocality. Those results then should be read as confirmations of superdeterminism rather than nonlocality. One is then free to propose a local hidden variable to explain quantum mechanics. Proposals like this have been made by Gerard 't Hooft,[12] among others.

Gerard 't Hooft is a truly great scientist, who in his twenties was singlehandedly responsible for a good portion of the key results that went into the construction of the standard model. I was very fortunate to take a course from him in graduate school, and I've always looked up to him personally. For many years he has been claiming to have constructed a deterministic and local hidden variable theory based on a cellular automaton, which is a model of a computer. If I understand correctly, it works for special cases; but he claims a more general validity based on an appeal to superdeterminism. But, details aside, between nonlocality and superdeterminism I am willing to bet that pursuing the former will bring us closer to the truth. I say this with some regret, as there are few theorists of his generation whom I admire more than Gerard 't Hooft.

GOING BEYOND PILOT WAVE THEORY AND COLLAPSE MODELS

The conclusion I come to is that none of the proposals for a realist quantum theory that I've presented so far are entirely compelling. Some are captivating, but none have either experimental support or the kind of elegance or completeness that can, for a time, substitute for that decisive experiment. So if you want to join Einstein, de Broglie, Schrödinger, Bohm, and Bell, and go beyond the statistical description of quantum theory to a description of beables that will tell us what exactly is happening in each individual quantum process, stay with us, for we are not yet done.

Are there lessons to take with us as we move beyond pilot wave theory and collapse models? Indeed there are. The most important lesson we can learn from the successes of the collapse models and pilot wave theory is that the wave function captures an element of physical reality. Let's see how this conclusion comes about.

The pilot wave theory asserts that everything in the universe has a dual existence—as a particle and as a wave. This solves the measurement problem because it keeps the particle. And it does so in a way that incorporates superposition, entanglement, and all their weird consequences because it keeps the wave. But is it right? I argued that impressive as it is, it has severe drawbacks. This brings us to our next option: to go beyond pilot wave theory to invent a new theory of beables.

Pilot wave theory succeeds because it posits that both particles and waves are real. But is this really necessary? Might there be a theory that accomplishes what pilot wave theory does which doesn't require the doubled ontology? This would also resolve the issue of the lack of reciprocity in the theory.

It would be extremely interesting if there were ways to reproduce the successes of pilot wave theory that had just one class of beables and not two. Waves or particles, but not both. Or something else entirely.

As a first try, we can ask what happens if we start with pilot wave theory and drop either the waves or the particles.

If we drop the particles, we no longer solve the measurement problem—unless we radically alter the behavior of the wave by hypothesizing spontaneous collapse. So dropping the particles leads back to either collapse models or the Many Worlds Interpretation.

We next try to drop the wave function, but keep the particles. What then is going to guide the particles? How are we to explain interference if we have only particles? Might we, for example, recover the wave function's guidance by giving the particles strange new properties?

Several physicists and mathematicians have tried to invent a theory of beables with just the particles, but none have succeeded. This is a long story, with some fascinating ins and outs, but the conclusion is simple: the wave function appears to capture an essential aspect of reality.[13] The closest to success I know of is an approach by the mathematician Edward Nelson called *stochastic quantum mechanics*. For many years I thought this was the right way, but then I understood it requires a large amount of fine-tuning to avoid instabilities.

This conclusion is upheld by a recent analysis by three specialists in quantum information theory, Matthew Pusey, Jonathan Barrett, and Terry Rudolph, who gave a new argument to the effect that the quantum state cannot be merely a representation of information an observer has about a system. It must be physically real, or represent something real.[14] So we seem to have only two choices: keep the wave function itself as a beable, as it is in pilot wave theory and collapse models, or find another beable that captures, in some different form, the physical reality which the wave function represents.

FOURTEEN

First, Principles

I went into physics hoping to contribute to solving the two great questions Einstein posed in his autobiographical notes: uniting quantum physics with spacetime and making sense of quantum physics.

Despite the efforts of many brilliant people over the more than half a century since Einstein wrote his autobiographical notes, these two problems remain unsolved. It is worth taking some time to ask why.

This question has been on my mind. Lately I find myself wondering if we have been going about completing Einstein's twin revolutions all wrong. We invent theories, such as loop quantum gravity, string theory, pilot wave theory, and others, but these do not go deep enough. Theories like these are models, which embody our ideas about nature, but they are not the deepest or purest expressions of those ideas.

Models exemplify ideas, but often in a simplified form, which allows the ideas' essential features and implications to shine through. The game Monopoly is a model of capitalism. Nonscientists often fail to appreciate how useful models can be—exactly because they are incomplete and leave things out—when one is in the stage of exploring the implications of an idea.

Ideas about nature are most purely expressed as either hypotheses or principles. A hypothesis is a simple assertion about nature, which is either true or false. "Matter is not infinitely divisible because it is made of atoms" is a hypothesis. So is "Light is a wave traveling through the electric and magnetic fields." Both of these hypotheses turned out to be true, but the history of science is littered with those that proved false.

A principle is a general requirement that restricts the form that a law of nature can take. "It is impossible to do any experiment that can determine an absolute sense of rest, or measure an absolute velocity" is a principle.

Einstein knew what he was doing when he introduced special relativity: he began his 1905 paper with two principles and deduced consequences directly from them. It is worth noting that the idea of unifying space and time into a single entity called spacetime was not part of Einstein's original conception of relativity. The idea of spacetime was introduced two years later by his teacher Hermann Minkowski as a model which exemplified Einstein's principles.

The problem with skipping the stage of principles and hypotheses and going right to models is that we can lose our way. It's easy to get trapped in a microscopic focus while trying to work out the details of those models. As Feynman once told me, "Make every question you ask in research a question about nature. Otherwise you can waste your life in working out the minutiae of theories that most likely will never have anything to do with nature." Even worse, we get caught up in petty competitions and academic turf battles between the adherents of different models.

Einstein expressed this lesson by insisting that we distinguish two kinds of theories. *Principle theories* are those that embody general principles. They restrict what is possible, but they don't suffice for the details. Those are supplied by the second kind of theory, which he called *constitutive theories*. These describe particular particles or specific forces that nature may or may not contain. Special relativity and thermodynamics are principle theories. Dirac's theory of the electron and Maxwell's theory of electromagnetism are constitutive theories.

So, my conclusion is that we need to back off from our models, postpone conjectures about constituents, and begin again by talking about principles.

Our strategy will then be to proceed to our goal of inventing a new fundamental theory in four steps: first, principles; second, hypotheses (which must satisfy the principles); third, models (which illustrate partial implications of the principles and hypotheses); then last, complete theories. Putting principles before theories raises an interesting question: Where do you find a language to state the principles, and a context to motivate and critique them? You don't want to use the language of existing theories because the whole point of the exercise is to get beyond them. Einstein would never have invented general relativity had he restricted himself to reasoning within the language of Newtonian physics.

Mathematics can sometimes provide new ideas and structures, and so is often a help. But new mathematics is usually not enough to invent new physics; otherwise Bernhard Riemann or William Kingdon Clifford would have invented general relativity. This is where a knowledge of philosophy can be the essential element, because a person with a philosophical education has in their toolbox a plethora of ideas and methods coming from the whole history of human beings' attempting to think about the fundamentals of our description of the world. And when it comes to basic questions like the nature of space and time, that history is rich with useful arguments and strategies to be tried out. So Einstein was not alone when he faced the need for new notions of space and time. It was as if he carried Galileo, Newton, Leibniz, Kant, and Mach in his back pocket and was able to converse with them and benefit from their insights. Similarly, a good knowledge of Plato, Kant, and others gave Heisenberg a language to go beyond Newtonian particles.

The twentieth century saw a flowering of philosophy in physics, which has further enriched the storehouse of useful ideas and arguments. Philosophy is indeed a living tradition, and if there was a time when philosophers of physics lagged behind in technical mastery of physics, that time is long over. So I will not apologize for going both to the sages of the past and to our contemporary philosophers to find language, contexts, and ideas to frame my search for new principles of physics.

Starting with principles has an immediate consequence, which is that we realize that quantum gravity and quantum foundations are different sides of a single problem. When physicists try to solve quantum gravity without regard to the problems of quantum foundations, and vice versa, we are taking the wrong approach. These two problems are deeply related. One reason is that because of quantum nonlocality, going beyond quantum mechanics means going beyond spacetime.

So I will proceed by putting forward principles which combine quantum phenomena with spacetime. After we have a good set of principles, the next step will be to frame hypotheses about how they are realized.

Our aim is to combine quantum physics and spacetime at the level of fundamental principles. I believe the right principles to shape this unification are the following:

PRINCIPLES FOR FUNDAMENTAL PHYSICS

1. Background independence.

A physical theory should not depend on structures which are fixed and which do not evolve dynamically in interaction with other quantities. This is a key concept, which takes some unpacking.

All physical theories to date depend on structures which are fixed in time and have no prior justification; they are simply assumed and imposed. One example is the geometry of space, in all theories prior to general relativity. In Newtonian physics, the geometry of space is simply fixed to be Euclidean three-dimensional geometry. It's arbitrary; it doesn't change in time, it can't be influenced by anything. Hence it is not subject to dynamical law.

In Newton's time, Euclid's was the only geometry known, so he had no alternative and didn't need to seek a justification for choosing it. But in the nineteenth century, Carl Friedrich Gauss, Nicholas Lobachevsky, and Riemann discovered an infinitude of alternate geometries. Any fundamental theory that comes after their work must justify the choice it makes for the geometry of space. The principle of background independence requires that the choice is made not by the theorist, but by the theory, dynamically, as a part of solving the laws of physics.

Non-dynamical, fixed structures define a frozen background against which the system we are interested in evolves. I would maintain that these frozen structures represent objects outside the system we are modeling, which influence the system but do not themselves change. (Or whose changes are too slow to be noticed.) Hence these fixed background structures are evidence that the theory in question is incomplete.

It follows that any theory with fixed external structures can be improved if the external elements can be unfrozen, made dynamical, and brought inside the circle of mutually interacting physical degrees of freedom. This was the strategy that led Einstein to general relativity. The geometry of space and time is frozen in Newtonian physics, and it is also frozen in special relativity. In these theories, the spacetime geometry provides an absolute and fixed background against which measurements are defined. General relativity unfreezes geometry, making it dynamical.

This is turning out to be a multistage process, because our theories have layers of frozen elements, which were laid down, like layers of sedimentation, during the long and complex history of our subject. General relativity unfreezes some aspects of geometry, but deeper structures, such as dimension and structures needed to define the continuous numbers or define a rate of change, remain frozen. So general relativity, beautiful as it is, cannot be the end of our search, and will require further completion.

Each step extends the range of the theory. It follows that the only complete theory of physics must be a cosmological theory, for the universe is the only system which has nothing outside of it. A theory of the whole universe must then be very different from theories of parts of the universe. It must have no fixed, frozen, timeless elements, as these refer to things outside the system described by the theory. It must be fully background independent.

This recognition that a cosmological theory cannot be achieved by just scaling up our current theories, but must be a radically new kind of theory, is the most important lesson learned so far in the search for a completion of Einstein's twin revolutions.*

It follows that quantum mechanics cannot be a theory of the whole universe because it too has fixed elements. These include the observables of the system and various relations they have, as well as the structure that gives rise to probabilities.†

This implies that there is no wave function of the universe, because there is no observer outside the universe who could measure it. The quantum state is, and must remain, a description of part of the universe.

We then seek to complete quantum theory by eliminating background structures. We do this by exposing and then unfreezing the background and giving it dynamics. In other words, rather than quantizing gravity we seek to gravitize the quantum. We mean by that the process of identifying and unfreezing those aspects of quantum theory which are arbitrary and fixed, making them subject to dynamical laws. Turning this around, we hope to understand the challenging features of quantum physics as consequences of separating the universe into two parts: the system we observe, and the rest, containing the observer and their measuring instruments.

Closely related to background independence is another key idea: *that the observables of physical theories should describe relationships.*

Leibniz, Mach, and Einstein taught us to distinguish absolute notions of space and time from relational notions. We say that location in space is absolute when there is a fixed meaning to where something is. A relative location is defined with reference to something else. Three blocks south of the supermarket is a relative

* For much more on this point, see my books, *Time Reborn*, and *The Singular Universe and the Reality of Time*, with Roberto Mangabeira Unger.

† In technical terms, the observables algebra and inner product.

location. Similarly, an absolute time is meaningful without reference to anything else, while relational time is always defined by its relation to another event or set of events.

This leads to our second principle:

2. Space and time are relational.

A relational observable, or property, is one that describes a relationship between two entities. In a theory without background structures, all properties that refer to location in space or time should be relational. Background-independent theories speak to us about nature through relational observables.

The third principle tells us nothing is left out.

3. Principle of causal completeness.

If a theory is complete, everything that happens in the universe has a cause, which is one or more prior events. It is never the case that the chain of causes traces back to something outside the universe.

Our next principle was introduced by Einstein, in his papers on general relativity.

4. Principle of reciprocity.

This principle states that if an object, A, acts on a second object, B, then B must also act back on A.

There is one more of these principles, and it is both subtle and powerful.

5. Principle of the identity of indiscernibles.

This states that any two objects that have exactly the same properties are in fact the same object.

Putting them in order, we have five closely related principles:

1. The principle of background independence
2. The principle that space and time are relational
3. The principle of causal completeness
4. The principle of reciprocity
5. The principle of the identity of indiscernibles

These are all aspects of a single principle, which Leibniz called the *principle of sufficient reason*. This states that every time we identify some aspect of the universe which seemingly might be different, we will discover, on further examination, a rational reason why it is so and not otherwise.

For example, given present knowledge, it seems that space might have more or less than three dimensions. (By this I mean the three large dimensions that we see around us; this doesn't count hypothetical tiny, "rolled-up" dimensions perceivable only on a subatomic scale.) This is because all our current theories, including general relativity and quantum mechanics, would also make sense in a world with a different number of spatial dimensions. Leibniz's principle of sufficient reason advises us that this must be because our current theories are incomplete. We must seek to complete our theories, and one sign of success will be when we find out why the number of large spatial dimensions is three.*

Leibniz believed we could uncover a rational explanation for every apparent choice God might seem to have made in the creation of the universe. He spoke of the state in which this understanding would be achieved as one of having "sufficient reason." His principle of sufficient reason states that the universe can be completely understood.

Each of the principles I've stated expresses this idea. For example, we could ask why the universe came into being where it was and not ten meters to the left. But everything would have happened just the same way, so this can't be a meaningful question. Therefore, absolute position is meaningless; only relative position is meaningful. A scientist who aspires to be rational must be a relationalist.

Our theories express these principles incompletely, but over time there has been a clear trend toward theories that explain more. Each time we explain a feature of the world in a way that limits the choice a creator might have had, we eliminate some of the arbitrariness we formerly perceived in the design of the world. As we understand the world better, it appears to us to be more rational. This happens each time we discover a hidden unity. A good example of this was Maxwell's discovery that light, electricity, and magnetism are not separate phenomena, but are different aspects of a single force. This discovery shows us that a world could not exist that has magnetism but no electric forces. And we understand that any world with electricity and magnetism must also have light.

I do not know if a complete understanding of nature will ever

* String theory does not do this; instead, it fixes the total number of dimensions, including possible microscopic dimensions. That could be a good thing, if it didn't give us infinite choices for the geometry and number of these hypothesized tiny dimensions.

be attained. But I do believe that our goal should be to always progress toward ever more complete understanding, which means we seek always less arbitrariness and more rationality. Hence I would propose we seek *ever more sufficient reason*.

I believe the progress of science is measured by such increases in our understanding of nature.

Special relativity is an improvement over Newtonian physics, and general relativity, by embracing a purely relational spacetime geometry, is an improvement on both. We can also say that quantum mechanics satisfies the principle of reciprocity better than Newtonian mechanics, but that pilot wave theory comes still closer to sufficient reason because it explains things quantum mechanics leaves unexplained, such as why individual events take place where and when they do.

But, as I've already mentioned, pilot wave theory fails to satisfy another of our principles: Einstein's principle of reciprocity. The pilot wave guides the particle, but the particle has no effect back on the wave. So we still have some distance to go.

The principle of sufficient reason advises us we can do better.

How shall we think of space and time in this new world of relations? Two chapters ago I drew a lesson from a survey of approaches to quantum foundations, which is that *space and time cannot both be fundamental*. Only one can be present at the deepest level of understanding; the other must be emergent and contingent. This seems ultimately to be forced on us by the nonlocality of entanglement, which leads to a tension between realist approaches to quantum mechanics and special relativity. The latter unifies space and time into spacetime, which the experimental tests of Bell's restriction suggest is transcended in individual quantum processes. I would then like to suggest that the tension is resolved by making one of the pair *space/time* fundamental, while the other is an emergent and approximate description, ultimately a kind of illusion. For many reasons, some described here, some the subject of earlier books,[1] I choose to focus on the hypothesis that time is fundamental, while space is emergent.

This is as far as principles take us. The next step is to frame hypotheses. I propose three hypotheses about what lies beyond spacetime and beyond the quantum:

Time, in the sense of causation, is fundamental. This means the process by which future events are produced from present events, called *causation*, is fundamental.

Time is irreversible. The process by which future events are created from present events can't go backward. Once an event has happened, it can't be made to un-happen.*

Space is emergent. There is no space, fundamentally. There are events and they cause other events, so there are causal relations. These events make up a network of relationships. Space arises as a coarse-grained and approximate description of the network of relationships between events.

This means that locality is emergent. Nonlocality must then also be emergent.

If locality is not absolute, if it is the contingent result of dynamics, it will have defects and exceptions. And indeed, this appears to be the case: how else are we to understand quantum nonlocality, particularly nonlocal entanglement? These, I would hypothesize, are remnants of the spaceless relations inherent in the primordial stage, before space emerges. Thus, by positing that space is emergent

* An event can be followed by a second event that reverses the action of the first, but then you have two events; this is not equivalent to a spacetime in which neither happened.

we gain a possibility of explaining quantum nonlocality as a consequence of defects which arise in that emergence.[2]

The combination of a fundamental time and an emergent space implies that there may be a fundamental simultaneity. At a deeper level, in which space disappears but time persists, a universal meaning can be given to the concept of *now*. If time is more fundamental than space, then during the primordial stage, in which space is dissolved into a network of relations, time is global and universal. Relationalism, in the form in which time is real and space is emergent, is the resolution of the conflict between realism and relativity.

Let's give a name to this version of relationalism, which emphasizes the reality and irreversibility of time and the fundamentality of the flow of present moments. Let's call it *temporal relationalism*. We can contrast it with *eternalist relationalism*, which investigates the hypothesis that space is fundamental, but time is emergent.

RELATIONAL HIDDEN VARIABLES

We thus seek a completion of quantum mechanics which is background independent and relational, and which is framed in a world where time is fundamental and space is emergent. If it involves hidden variables, these must express relations between particles. Thus, the hidden variables do not give us a more complete description of an individual electron; they must describe relations which hold between one electron and other electrons. We can call these *relational hidden variables*.

Indeed, what is more relational than the deepest and subtlest of the quantum mysteries, which is entanglement? A relational formulation of quantum physics will start by putting entanglement first. If, as we hypothesized, space is emergent, distance in space must be derivative of more fundamental relations. Perhaps this more fundamental relation, from which space emerges, is entanglement.*

The hidden variables in pilot wave theory are the trajectories of the particles. They are not relational; they do in fact just give us more information about each of the particles, individually. However, there is already a large dose of relationalism in pilot wave theory. This is inherent in the fact that for a system of more than one particle, the wave function lives not in ordinary space, but in the space of configurations of the total system, which consists of several particles. This is, as I explained in chapter 8, necessary to incorporate entanglement.

I first formulated the concept of a relational hidden variable theory, including the hypothesis that space is derivative of more fundamental relations, especially entanglement, early in my career. I wrote up[3] a formulation of a relational hidden variable theory in 1983; this was the first of several such efforts.[4]

My 1983 theory was based on a simple idea. Suppose you have a system of particles in space. In an absolute description, you code in the location of each particle individually by giving coordinates in space. These coordinates are absolute; they refer to an observer outside the system—for Newton this was God himself. In a relational description you could use only the relative distances between each pair of particles. These no longer depend on reference to an observer outside the system.

There is a relative distance between every pair of particles. Hence, the relative distances can be represented as a table of numbers. The entry "10 down and 47 over" gives the distance between

* This is not a new idea; as I noted in chapter 9, Roger Penrose mentioned it as motivation for his spin networks model in the early 1960s.

the 10th and 47th particles. Another name for such a table of numbers is a matrix. In my relational hidden variable theory, the hidden variables were such a matrix. My 1983 theory utilized a large matrix of complex numbers to describe a system of many particles living in a two-dimensional space. When the number of particles was large, the probabilities for the motion of the particles were approximately described by Schrödinger's equation.

$$\begin{bmatrix} 6 & 8 & 1 & 5 \\ 1 & 10 & 16 & 3 \\ 21 & 0 & 9 & 61 \\ 4 & 15 & 201 & 7 \end{bmatrix}$$

FIGURE 11. A matrix is a table of numbers, made up of rows of columns.

There are by now several proposals that go beyond the quantum by starting with pilot wave theory and trading in the wave function for a deeper structure described in terms of matrices. Relational hidden variables theories based on matrices have also been proposed by Stephen Adler[5] and Artem Starodubtsev.[6]

A matrix assigns a number to every pair of particles. Another structure that does so is a graph, which is a simple structure built of points, connected by lines. Each pair of points is either connected by a line or not. We can assign a one to the pair if they are connected and a zero if they are not, and then we have a matrix representing the same structure.

Graphs and matrices are thus both ways to express the hypothesis that the fundamental beables underlying physics are a network of relations. These relations may express quantum entanglement and nonlocality.

There is no purer model of a system of relations than a graph or network. Interestingly enough, networks are ubiquitous in those approaches to quantum gravity which are in accord with the principle of background independence. These include loop quantum gravity, causal sets, and causal dynamical relations. This suggests two exciting deepenings of our hypotheses: First, space emerges from the fundamental network. Second, quantum physics arises from nonlocal interactions left over when space emerges.

However, networks fit uneasily into space, if "nearby" in the emergent space is to correspond with "nearby" in the network. The reason is simple: consider two points in the graph, each corresponding to a point in the emergent space. Suppose they are far away from each other in space and also far away on the graph. But now add a link to the graph directly connecting those two points. All of a sudden they are neighbors on the graph, but still far away from each other when considered in terms of the emergent space.

In our work with Fotini Markopoulou, we called such connections *defects of locality*. They look like narrow wormholes. We showed that they will be common in loop quantum gravity.[7] This led us to another paper where we derived quantum mechanics from averaging over the nonlocal interactions which might arise from such defects of locality.[8] A bit tongue in cheek, we called this "Quantum Theory from Quantum Gravity."*

* Juan Maldacena and Leonard Susskind have since introduced a version of this idea they call ER=EPR, where ER stands for an Einstein-Rosen bridge, which is a wormhole connecting two points far from each other in space ("Cool Horizons for Entangled Black Holes," arXiv:1306.0533).

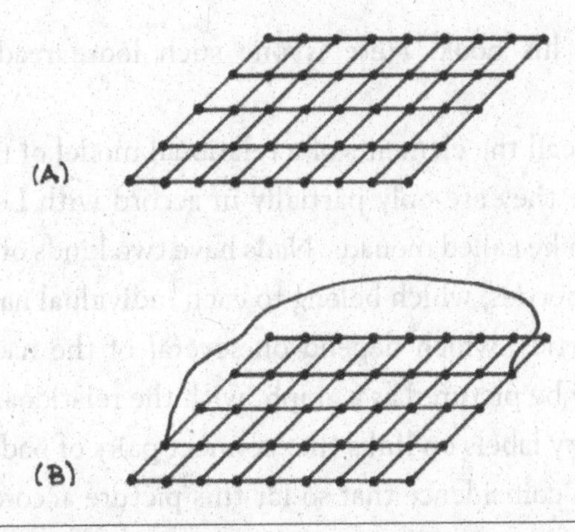

FIGURE 12. DISORDERED LOCALITY (A) A lattice of points, embedded in space, which we call local because points which are far away in terms of steps on the lattice are far away in the space it is embedded in. (B) By adding a new link that connects far away points, we disrupt locality because the connected points are still far away in space, but are only one step apart on the lattice.

I MET RICHARD FEYNMAN ONLY a few times, but on two occasions he was kind enough to ask about my work. Each time he responded the same way. He listened carefully and then suggested that the idea I described to him wasn't crazy enough to have a chance to work. What I believe he meant by that was that my idea didn't go deep enough. In any case, that is how I feel about my earlier attempts to make a relational hidden variable theory based on matrices and networks. They solve the problem of giving a completion of quantum mechanics at a technical level, but in other aspects they come up short. One way to tell is that the Schrödinger equation only comes out as a prediction of the theory if you hammer out the imperfections and fine-tune the equations.

To go deeper into the relational idea, we can go back to Leibniz for inspiration. Leibniz sketched a purely relational view of the universe in a short book, *The Monadology,*[9] written in 1714. Since we are interested only in getting inspiration from Leibniz, we don't care to accurately reproduce his vision. We are free to creatively misinterpret his book. Here is one such loose reading of *The Monadology*.

We shall call the elements of a relational model of the universe *nads* because they are only partially in accord with Leibniz's elements, which he called monads. Nads have two kinds of properties: intrinsic properties, which belong to each individual nad, and relational properties, which depend on several of the nads. A nadic universe may be pictured as a graph, with the relational properties represented by labels on links that connect pairs of nads.

It is not a coincidence that so far this picture accords with the description of the world given in loop quantum gravity. There, a state of the world is described by a graph with labels on it.

Each nad has a *view of the universe*, which summarizes its relations with the rest. One way to talk about the views is in terms of neighborhoods (or zones) of the graph. Let's talk about the view of a nad called Sam. Consider the nads one step away in the graph from Sam: they are the first, or nearest, neighbors. The first neighborhood consists of Sam and her nearest neighbors, together with the relations they share, which are indicated on the links between them.

To construct Sam's second neighborhood, add in the nads two steps away from her, and all their relations with each other and with their neighbors who are one step away (who are also included). And so on. These neighborhoods constitute Sam's views of her universe.

We can compare Sam's views to the views of another nad—let's call him Sue. Sam and Sue have identical first and second neighborhoods, which is to say, we couldn't tell them apart if we can only see that far.

But let us posit that our relational, nadic universe obeys Leibniz's principle of the identity of indiscernibles. Then Sam's and Sue's neighborhoods must differ at some point; otherwise they would have identical views, which is forbidden by that principle.

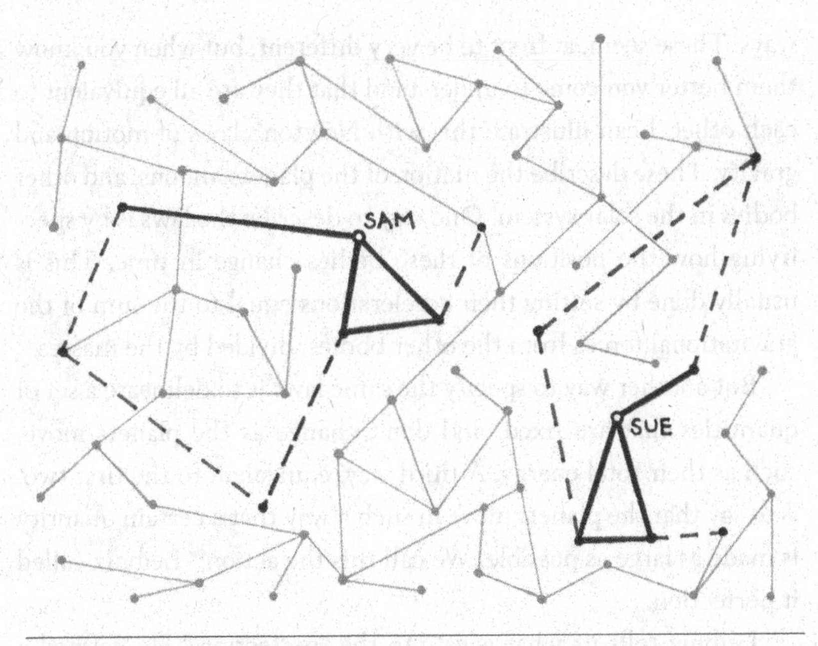

FIGURE 13. The first and second neighborhoods of Sam and Sue, defined by the connectivity of the graph they inhabit, are identical, but the third and higher neighborhoods distinguish them.

This implies there must be some number of steps at which the two neighborhoods differ. We call that number the *distinction* of Sue and Sam.

Leibniz posited that the actual universe is distinguished from many possible universes by "having as much perfection as possible." If we strip this of its poetic or allegorical meaning, what Leibniz is doing is positing that there is some observable quantity which is larger in the real universe than in all the other possible universes. This is shockingly modern, as it anticipates a method for formulating laws of nature that was developed later and only came into fruition during the twentieth century. The quantity that is maximized, which Leibniz called "perfection," we call an *action*.

Feynman liked to emphasize that a beautiful feature the laws of physics enjoy is that they can be formulated in several different ways. These seem, at first, to be very different, but when you know them better you come to understand that they are all equivalent to each other. I can illustrate this with Newton's laws of motion and gravity. These describe the motion of the planets, moons, and other bodies in the solar system. One way to describe the laws is by specifying how the positions of these bodies change in time. This is usually done by setting their accelerations equal to the sum of the gravitational forces from the other bodies, divided by the masses.

But another way to specify the same laws is to delineate a set of quantities that are fixed, and don't change as the planets move, such as their total energy. A third way, equivalent to the first two, is to say that the planets move in such a way that a certain quantity is made as large as possible. We call this the action;* Leibniz called it perfection.

Leibniz tells us what goes into the "perfection." He defined a world with "as much perfection as possible" to be one having "as much variety as possible, but with the greatest order possible."

What does Leibniz mean here by "variety"? I believe that Leibniz meant that the views of the different monads should differ as much as possible. So by maximizing perfection, Leibniz means we should maximize the variety of different views.

Inspired by this picture, Julian Barbour and I constructed numerical measures of the variety inherent in a system of relations.[10] We noticed that as variety increases, less information is needed to pick out and distinguish each view from the others. That is, everything else being equal, we prefer worlds where any pair of nads has neighborhoods which differ at a small number of steps.

For Leibniz, sufficient reason had to be founded on a notion of maximal perfection.

* More precisely, the negative of the action.

> And this [sufficient] reason can be found only in the fitness, or in the degrees of perfection, that these worlds possess. . . . This interconnection (or accommodation) of all created things to each other, and each to all the others, brings it about that each simple substance has relations that express all the others, and consequently, that each simple substance is a perpetual, living mirror of the universe.

He then reaches for a metaphor to describe this, and comes up with the different views of a city.

> Just as the same city viewed from different directions appears entirely different, and, as it were, multiplied perspectively, in just the same way it happens that, because of the infinite multitude of simple substances, there are, as it were, just as many different universes, which are, nevertheless, only perspectives on a single one.[11]

This is indeed a metaphor that Jane Jacobs would have appreciated, as it captures a notion of urban diversity championed by her and embraced by philosophers of the city, such as Richard Florida, since.

This urban metaphor inspires a hypothesis about how space and locality break down. If you stand next to me and we both look out, by virtue of our proximity we have similar views of the rest of the universe. Our views cannot be identical, because we cannot coincide, by virtue of both the Pauli exclusion principle and the identity of indiscernibles. But the closer to each other we stand, the more similar are our views.

Because we are close to each other, we can interact easily, and indeed, the closer we stand the higher is the probability that we interact through an interchange of quanta such as photons. This is basically what we mean when we say physical interactions are local.

But suppose we have this backward. What if we interact with high probability exactly because our views are similar? Suppose that the probability for us to interact increases with the increasing similarity of our views, and decreases if our views begin to differ.

If this is right, then the fundamental relation determining how often we interact is how similar our views are—and distance in space is derivative from that.

Now, for big, clunky things like ourselves, made up of vast numbers of atoms, this is as far as it goes. But consider what it takes for atoms to have similar views. Atoms have many fewer degrees of freedom, hence fewer relational properties. So atoms which are far away from each other in space may still have similar neighborhoods, just because there are vastly fewer configurations their local neighborhoods could take. This suggests that perhaps similar atoms, with the same constituents and similar surroundings, interact with each other just because they have similar views.

These interactions would be highly, highly nonlocal. But in my recent work, I have showed that this could be the basis of quantum physics.[12]

Consider a hydrogen atom in a water molecule dancing in the air in front of me. This has a first neighborhood consisting of an oxygen atom, and a second neighborhood consisting of the whole molecule. The same is true of every hydrogen atom in a water molecule everywhere in the universe. So I am going to trust my relational instincts and take the crazy step of positing that all these atoms are interacting with each other, just because their views are similar. More specifically, I will posit that the interactions act to increase the differences between these atoms' views. This will go on until the system has maximized the variety of views the atoms have of the universe.

In a recent paper, I showed that the hypothesis of maximal variety leads to the Schrödinger equation, and hence to quantum mechanics. This happens because there turns out to be a mathematical similarity between the variety and Bohm's quantum force. As a result, Bohm's quantum force acts to increase the variety of a system. It does so by making the neighborhoods of all the different particles as different from each other as possible.

In this approach the probabilities in quantum mechanics refer to an ensemble that really exists, the ensemble of all systems with similar views. This is a real ensemble, in that the elements are not located in our imagination; they are, each and every one, a part of the natural world. This is in accord with the principles of causal completeness and reciprocity.

This was the basis of a relational hidden variable theory I proposed, which I called the *real ensemble formulation of quantum mechanics*. From it, I could derive the Schrödinger formulation of quantum mechanics from a principle that maximizes the variety present in real ensembles of systems with similar views of the universe.

On the technical side, this theory borrows from the many interacting classical universes theory I described in the last chapter, only the ensemble of similar systems does not come from other universes parallel to our own; instead they are similar systems far away in distant regions of our own single universe.

In this theory, the phenomena of quantum physics arise from a continual interplay between the similar systems that make up an ensemble. The partners of an atom in my glass of water are spread through the universe. The indeterminism and uncertainties of quantum physics arise from the fact that we cannot control or observe those different systems. In this picture, an atom is quantum because it has many nearly identical copies of itself, spread through the universe.

An atom with its neighborhood has many copies because it is close to the smallest possible scale. It is simple to describe, as it has few degrees of freedom. In a big universe it will have many near copies.

Large, macroscopic systems such as cats, machines, or ourselves have, by contrast, a vast complexity, which takes a great deal of information to describe. Even in a very big universe, such systems have no close or exact copies. Hence, cats and machines and you and I are not part of any ensemble. We are singletons, with nothing similar enough to interact with through the nonlocal interactions. Hence we do not experience quantum randomness. This is a solution to the measurement problem.

This theory is new, and, as is the case with any new theory, it is most likely wrong. One good thing about it is that it will most likely be possible to test it against experiment. It is based on the idea that systems with a great many copies in the universe behave according to quantum mechanics, because they are continually randomized by nonlocal interactions with their copies.

I argued that large complex systems have no copies, and hence are not subject to quantum randomness. But can we produce microscopic systems, made from a small number of atoms, which also have no copies anywhere in the universe? Such systems would not obey quantum mechanics, in spite of being microscopic.

We have the capability to do just that using the tools of quantum information theory. Indeed, a sufficiently large quantum computer should be able to produce states involving enough entangled qubits that they are very unlikely to have any natural copies anywhere in the observable universe. This suggests that the real ensemble theory can be falsified by making a large quantum computer that works exactly as predicted by quantum mechanics.

Science progresses when we invent falsifiable theories, even if the result is that they get falsified. It is when theorists invent non-falsifiable theories that science gets stuck.

And what about systems with small numbers of copies? These behave neither quantum mechanically nor deterministically. They will have to exhibit behavior of a new kind which is neither classical nor quantum. This will give us further opportunities to test this new theory.*

THE PRINCIPLE OF PRECEDENCE

The real ensemble theory depends on a system being able to recognize and interact with other systems which are similar to it, in the sense that they have a similar view of the universe of relations, no matter where they are in the universe. According to this hypothesis, similarity or difference of views is more fundamental than space; space emerges to describe the rough order created by similarity of views. Two systems may interact if their views are similar enough. Often that reflects their being nearby in space and time,

* In this real ensemble formulation, the information in a wave function of a quantum system is spread throughout the universe, coded into the configurations of the copies. A key question is how many copies a system must have for the information coded into the copies to be adequate to reproduce the information in the wave function. That information increases exponentially with the number of particles in the quantum system. But the number of copies of a system the universe will likely contain decreases rapidly with the number of particles that make up the system. So there is a size of a system beyond which the information in the copies does not suffice, with the consequence that either quantum mechanics breaks down, or this approach is wrong. I suspect that even modest quantum computers will cross this line.

but not always, and it is the latter cases that underlie quantum phenomena.

What happens if we apply this viewpoint to systems at different times? Might a system interact with systems in the past that have similar views? If this is possible, we can use the influence of the past on the present to find a new understanding of what the laws of nature are. This leads to a novel idea, which I call the *principle of precedence*.[13]

To explain it in simple terms, it helps to use operational terminology, in which a quantum process is defined by three steps. The first is its preparation, which picks the initial state. Next we have an evolution, during which it changes in time according to Rule 1. At the end we have a measurement, which is governed by Rule 2. We have several choices about what we measure, but whichever we choose, several different outcomes are possible. Quantum mechanics predicts that the probabilities for these different outcomes will depend on the preparation, the evolution, and the choice of what we measure. If we know the forces acting on the system during the evolution, we can use Rules 1 and 2 to predict the probabilities of the different outcomes.

It is common to believe that, once the environment of the system is fixed, Rule 1 evolves the system in time as dictated by the fundamental laws. These laws are presumed not to change in time. As a consequence, we can say the following. For every quantum system we study in the present, defined by a specific preparation, evolution, and measurement, there will be a collection of similar systems in the past. These are similar in the sense that they had the same preparation, evolution, and measurement as our present system. Now, the fact that the laws don't change implies that the probabilities for different outcomes also don't change.

As a result we can say that

*The probabilities for different outcomes to result in the present experiment are the same as if we picked random outcomes from the collection of past similar instances.**

We can call this the *law of precedents.*

Now I would like to make a simple but radical proposal. The law of precedents is usually understood to be a consequence of the existence of unchanging laws. But actually, this law of precedents is all we need of law. We can posit that there is no law except the law of precedents. Instead of the above, we posit that

The probabilities for different outcomes to result in the present experiment are arrived at by picking random outcomes from the collection of past similar instances.

By this I postulate that a physical system has access to the outcomes of systems with similar preparations, evolutions, and measurements in its past (we call these "similar systems," for short). Our hypothesis is then

A physical system, when faced with a choice of outcomes of a measurement, will pick a random outcome from the collection of similar systems in the past.

This law of precedents guarantees that most of the time, the present will resemble the past, in that the probabilities for the

* I.e., those with the same preparation, evolution, and measurement.

various possible outcomes of the same experiment will be unchanged.

If this is right, the appearance that atoms are governed by unchanging laws is an illusion created by the fact that the universe is old enough and big enough that there is ample precedent for most situations an atom will find itself in.

But what if there are no precedents? What if we prepare a quantum state which has never so far existed in the history of the universe? If we make a measurement of it, how will we determine its outcome, if there are no past similar instances to refer to?

I don't know the answer to this question. This could be and, I hope, will be a question for experimental physics. The standard belief in a timeless fundamental law has no problem making a prediction, by applying the known law to the new situation. If the experiments always confirm that answer, we can deduce that the principle of precedence is wrong. However, if precedence is the key to lawfulness, then the response to a novel situation, a novel quantum state, will be novel.

After many repetitions precedence builds up, and there will no longer be surprises. The transition, though, from novelty to precedence should be open to experimental investigation.

The site for such investigations is again likely to be laboratories where experimentalists are preparing entangled states of several atoms. Such states will at some point soon be complex enough that it would be safe to deduce they have no precedents in the history of the universe. So very soon it ought to become possible to test the principle of precedence experimentally, and perhaps discover the process by which precedence builds up.

FIFTEEN

A Causal Theory of Views

Each of us theorists has his or her commitments: the guesses about nature you are willing to bet your career on. Personally, I am a realist, a relationalist, and, indeed, a temporal relationalist. I believe that quantum mechanics is incomplete and aim to construct a realist theory according to the principles of temporal relationalism, which can stand as a simultaneous completion of quantum mechanics and general relativity. I have hopes that this theory will not only resolve the puzzles in the foundations of quantum theory, but will lead to the discovery of the right quantum theory of gravity, as well as address mysteries in cosmology and particle physics coming from the universe's apparent freedom to choose both laws and initial conditions.

In this closing chapter I'd like to describe one path we might take to reach this goal, and then tell you about some very recent work that brings us a few steps along this path.

This is a theory of nads, of the sort I've been describing, with two additional ideas. First, we take seriously Leibniz's idea that what is real in a purely relational description of the world is the views that each nad has of the rest of the universe. The views don't represent what is real; they are what is real. This means that the views themselves are the dynamical degrees of freedom, the protagonists of our story. This indeed brings our nads closer to what Leibniz called monads (although there are still some differences).

But to what exactly do the nads correspond in the world we are familiar with, and of what do their views consist?

If we want a correspondence with general relativity, it is natural to presume that the nads are events. In relativity theory, events are things that happen at a single place and time. They are fundamental to general relativity's picture of the world. You can think of them as moments when something changes at one place: for example, two particles colliding make an event. A world made of events is a world in which "to become" is more fundamental than "to be."

If the nads are events, what do the relations between them describe? The short answer is causation. Events cause other events.

Each event is woven into the history of the universe through relations with the other events, which express which events might be a cause of which. These *causal relations* chart the history of processes of change.

We can extract how these relations work from general relativity. Given that causes can propagate only at the speed of light or less, we say that an event B is in the *causal past* of another event, A, if a physical cause could have traveled at the speed of light or less from B to A. If this relation holds, then conditions at B might have contributed to causing conditions at A.

Under the same condition we also say that A is in the *causal future* of B.

Given any two events, A and B, we usually require of general relativity that only one of the following three things must be true.

Either A is in the causal future of B, or B is in the causal future of A, or they are causally unrelated because no signal traveling at the speed of light or less could have passed between them. This rules out closed causal loops in which A is in both the causal future and causal past of B. Exotic histories with closed causal loops are fun to speculate about, but they raise puzzles and paradoxes. I see no reason to presume closed causal loops are part of nature, especially as I want to presume that causation is fundamental, and fundamentally irreversible.*

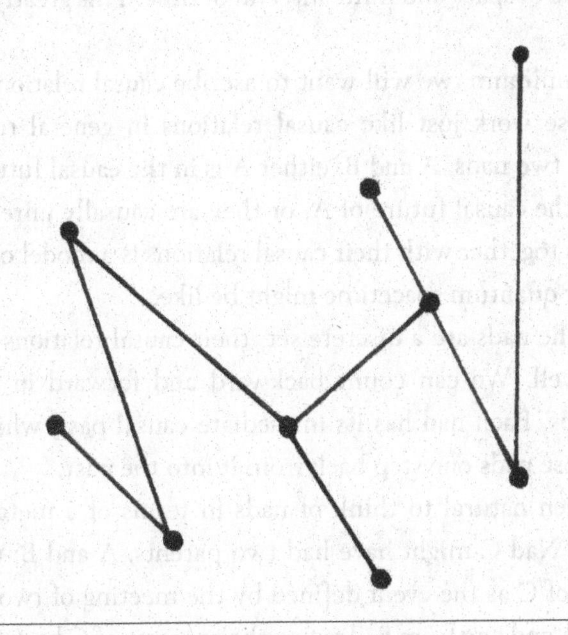

FIGURE 14. A set of discrete events, connected by causal links.

* Some relativists point to the existence of solutions to the Einstein equations which have closed causal loops. I don't think this has any force because the universe is described by at most one solution to general relativity, and that solution need not have every exotic behavior shown in other solutions. More definitively, those solutions which have closed causal loops (including one proposed by the great logician Kurt Gödel) are very special in that they have a lot of symmetry. If we impose the principle of the identity of indiscernibles, then solutions with symmetries are excluded. These solutions are also unstable and collapse to singularities at the faintest hint of a perturbation.

If we say what the causal relations are between every pair of events, we are describing the universe in terms of its *causal structure.*

According to general relativity, spacetime consists of a continuous infinity of events. Instead, I follow some of the pioneers of quantum gravity, who hypothesize the nads to be a discrete set of fundamental events. Discrete means they can be counted, whether the count is finite or infinite. We will also require that even if their total numbers are infinite, there is a finite number within any finite volume of space and finite interval of time. This greatly simplifies things.

At a minimum, we will want to ascribe causal relations to the nads. These work just like causal relations in general relativity. Given any two nads, A and B, either A is in the causal future of B, or B is in the causal future of A, or they are causally unrelated. A set of nads together with their causal relations is a model of what a discrete or quantum spacetime might be like.

Since the nads are a discrete set, their causal relations are discrete as well. We can count backward and forward in discrete causal steps. Each nad has its immediate causal past, which consists of those nads one step back from it into the past.

It is then natural to think of nads in terms of a metaphor of parentage. Nad C might have had two parents, A and B; then we can think of C as the event defined by the meeting of two causes, one from A and one from B. Tracing the ancestry of C back through A and B to their parents, and beyond, gives us a network of causes stretching deep into the past. C in turn might have two progeny, D and E, which it influences.

At this point, we have in front of us a possibility of breathtaking simplicity. We can suppose that the events which make up the history of the world have fundamentally only these causal relations.

All other entities and all other properties in nature are to be derived from a large but discrete set of events whose only property is which causes which. This radical suggestion was made by Rafael Sorkin,[1] and developed in close collaboration with a group of friends and enthusiasts. They call it the *causal set theory*.

A causal set is simply a discrete set on which there are defined only causal relations, satisfying the condition that an event is never its own cause. One also requires that given any two events A and B, only a finite number of events are in both the causal future of B and the causal past of A.

I admire the ambition and radical purity of causal set theory. It is a completely relational description of spacetime, in which each event is defined completely in terms of its place in the network of causal relations.

One very good feature is that the geometry of a spacetime can, to a good approximation, be captured by a causal set. This is done by a method analogous to how polls of our political views are taken. Rather than ask everyone's views, those of a small, randomly chosen sample are queried. Similarly, one can pick out a random sample of events in a spacetime and record their causal relations with each other. One loses a lot of information, but if one picks an event per some fixed volume of space and unit of time, one gets a representation of the causal relations which is accurate down to that scale.

However, Sorkin and his collaborators hypothesize that the reverse is also the case. They believe the history of the universe is, at its most fundamental, a discrete causal set, from which emerges, on a sufficiently large scale, the illusion of a continuous spacetime. Just like a liquid appears to us continuous but is actually made up of discrete atoms, the events of the causal set would constitute the atoms of spacetime.

One great success of the causal set theory is that it predicted the rough value of the cosmological constant. Sorkin derived this prediction before the cosmological constant was measured.[2] It was the only approach to quantum gravity to do so.

The causal set hypothesis is one of several competing hypotheses concerning the properties of spacetime atoms. Compared to the others, such as spin foam models, it enjoys the great advantage of its utter simplicity, in that the only properties of events are their causal relations. This greatly narrows down the possible forms that a fundamental law of spacetime atoms could take.

This radical simplicity is also behind a very formidable obstacle that this approach faces, which is called the *inverse problem*. As I said earlier, given a continuous spacetime, we can easily sample its events to find a causal set. But the reverse is almost never the case. In the world of possible causal sets, almost none provide an approximate description of a spacetime, with three dimensions of space. This makes it seem as if there is more to a spacetime than a rough description of a network of causal relations.

Quantum gravity, or the problem of understanding spacetime within quantum theory, has certainly proved to be a formidable challenge. It helps to put the challenge of discovering the atoms of spacetime in perspective by comparing it with the history of the hypothesis that matter is made of atoms.

In the case of matter, the challenges facing atomists in the nineteenth and early twentieth centuries were twofold. First, they needed to discover the fundamental laws that govern the atoms. Second, they had to deduce from those fundamental laws the rough properties we perceive matter to have. They had to understand how the illusions of solids, liquids, and gases arise as consequences of the more fundamental atomic laws. Theorists of quantum gravity face the same two challenges.

We should be mindful of two lessons from the history of the original atomic hypothesis. The first is that progress on the first challenge—that of discovering the laws of atomic physics—didn't begin to be made until we had experiments that could verify that atoms really existed and reveal to us some of their properties.

History also teaches us that the second challenge—deriving the bulk properties of the various phases of matter—may be easier to address than the first. Half a century before we began to make real progress on uncovering the laws of atomic physics, a few pioneers had already made substantial progress on the second challenge. The reason is that the behavior of matter in bulk turns out not to depend very much on the details of atomic physics. One needed to know only that there are atoms and that they interact by forces that are short-range (i.e., could only act over a short distance).

This lesson is taken to heart by some quantum gravity theorists who seek to derive the law that governs spacetime on a macroscopic scale, namely general relativity, from simple hypotheses about the atoms of spacetime. This direction was pioneered by Ted Jacobson,[3] and it has succeeded to a significant degree. This makes it likely that the known laws of physics, which operate on scales we can observe—much larger than the fundamental Planck scale—don't depend very much on the laws that govern the atoms of spacetime.

This is bad because it means that the known laws hold few clues which might help reveal the truly fundamental laws. Indeed, there are, it would seem, just two clues. The first has to do with how information flows through spacetime, and is the following: To derive general relativity from the properties of hypothetical atoms of spacetime, one must posit that there is a maximum rate at which information may flow through a surface in space. This rate of information flow cannot be greater than the area of that surface, when counted up in fundamental Planck units.* This is called the (weak[4]) *holographic hypothesis*.†

If this holographic hypothesis is fundamental, then it has to make sense to speak of a flow of information all the way down at the tiny scales where quantum gravity operates. But information is influence, as is expressed by defining it as the distinction that makes a difference. So a flow of information defines (or depends on) a causal structure. Thus the holographic hypothesis requires that we must have a causal structure to guide, or express, the flows of information. This is one reason to believe that causal structure is fundamental.

The second clue is that to derive general relativity following Jacobson's argument, we have to keep track of the flows of energy through the same surfaces. This suggests that energy is a fundamental quantity that makes sense all the way down to the level of the fundamental events. The great insight of Jacobson is then to have realized that, most fundamentally, the equations of general relativity encode a relationship between flows of energy and flows of information, both flows being guided by the causal structure.

Because of the first clue, I favor the hypothesis that the history of the universe contains a set of events and their causal relations, i.e., that the universe is a causal set. But, because of the inverse problem, I do not believe the radical hypothesis that the only properties events enjoy are causal relations. I am willing to believe that the causal relations are the only relational properties needed, but I believe there must be further properties, which are intrinsic to the events. The second clue leads me to posit that among these

* These fundamental units of area are equal to the product of Newton's gravitational constant and Planck's constant.

† For more about the holographic hypothesis, see my book *Three Roads to Quantum Gravity*.

intrinsic properties, events are endowed with energy, which flows between them, following the causal relations.

I then would propose that each event has a certain quantity of energy, and that energy is transmitted from past events to future events along the causal relations. An event's energy is the sum of the energies received from the events in its immediate causal past. That energy is divided up and transmitted to the events in its immediate causal future. In this way the law of conservation of energy, according to which energy is never created or destroyed, is respected.

Special relativity tells us that energy is unified with momentum, so I would have momentum propagated from past events to future events as well. In collaboration with Marina Cortês, I invented a causal set model which incorporates flows of energy and momentum, which we call an *energetic causal set.*[5]

The history of the universe, according to an energetic causal set model, consists of events which are each the causes of future events, to which they transfer some energy and momentum. But there is no spacetime, fundamentally; there is just the discrete set of events connected by causal relations, with the events and the relations endowed with energy and momentum.

One early success of this approach was a solution to the inverse problem. At least in simple cases in which space and time each have one dimension, we were able to derive the emergence of spacetime directly from the energetic causal set models.

IT IS TIME we talked about energy.

Each of the major physical theories, from Newtonian physics down through general relativity and on to quantum field theory, has equations of motion that tell how some entity changes in time.

For Newton, that entity is the position of a particle, while for quantum field theory it is the value of a field at every point in space. It is highly significant that all these equations of motion share a common structure. There is a configuration variable—the positions of the particles or the values of the fields. Then there are certain additional, dynamical quantities, which are called so because they come into the laws that tell how the particles move around or how the fields oscillate. The most important of these are momentum and energy.

Each particle carries a certain quantity of energy and momentum. When two particles interact, they exchange some of their energy and momentum. One may gain a bit, while the other loses, so long as the total energy and total momentum are conserved.

The structure of these theories is always the same: there are two fundamental equations. The first tells how the positions of the particles change in time, in a way that depends on the particles' momentum.* The second equation tells how the momentum changes in time, and this depends on the particles' positions. So the two quantities, position and momentum, are intertwined; the change of one depends on the other. We say that two quantities, related in this way, are dual. Position and momentum are dual. So are the electric and magnetic fields.

I believe that the fact that this pattern of dual equations is universal in physics is a deep property of nature. It is also restricted to physics. Other sciences describe systems that change in time, such as computers or ecosystems or markets or organisms. They each have their equations. But in none of these cases do the equations have this dual structure involving configuration variables, momenta, and energy, the latter two of which are conserved in total.

* In the Newtonian case, the momentum of a particle is proportional to its velocity. The constant of proportionality is the mass.

This is one reason I don't think it's very helpful to imagine that the physical universe is a computer.

The conservation of momentum is important for another reason. It explains the principle of inertia, which is the deepest principle of physics so far posited.

Why is there this duality, involving configuration and momentum variables? Why is the world such that energy and momentum are conserved? There is an old answer to these questions, which is based on a deep theorem of Emmy Noether, which she proved in 1915. It involves the notion of symmetry, which is a transformation that changes a system in some way that doesn't change the laws of motion. Rotations are symmetries, as are translations in space and time; so long as the entire system is rotated or translated together, then the laws of motion are unaffected. Noether's theorem states that *for every symmetry in nature that is based on a transformation that varies continuously, there is a conserved quantity*. Symmetry in space implies that momentum is conserved. Symmetry in time explains the conservation of energy.*

This suggests that space is fundamental, while energy and momentum are emergent properties of space, reflecting its symmetries. This is a standard view, but I believe the reverse is closer to the truth.

While Noether's theorem reflects a true insight, it cannot apply to a fundamental theory. This is because we require that the fundamental theory satisfy the principle of the identity of indiscernibles. But that principle implies that there are no symmetries in nature. Think of a body that is invariant under a rotation, such as a sphere or a cylinder. The fact that it is symmetrical means that it is unchanged by a rotation. That is, an observer cannot tell the

* And rotational symmetry implies the conservation of angular momentum.

difference between the body before and after it is rotated. But this is true because there are on the body circles of points, which are all identical to each other. Similarly, an infinite straight line is invariant under a translation along its length because under such a translation, each point is taken to another point with identical properties. In each case we see that the existence of a symmetry means there are distinct points with identical properties, which violates the principle.

Symmetries are properties of fixed backgrounds, and the occurrence of a symmetry in a theory is a clear sign that that theory is background dependent. A symmetry is an operation that translates or rotates the system we are studying, with respect to the background, which is left unchanged. Symmetries characterize a system that has been isolated from a larger universe, and arise from what is ignored in that isolation.

We have posited that the fundamental theory is background independent, which means there are no symmetries. This in turn means that we cannot regard energy and momentum, and their conservation, as emergent from the properties of space. But we still have to explain why energy and momentum play the ubiquitous role they do in the structure of the equations of physics.

Further, we have hypothesized that space is not present at the fundamental level in nature, but is emergent. So if we want energy and momentum to play a role in physics, there seems to be no alternative but to put them in at the beginning.

What we want is an inverse of Noether's theorem, which assumes that energy and momentum and their conservation are fundamental, and tells us the conditions under which space may emerge as an approximate description of subsystems of the whole.

So we are left with a picture in which causal relations, energy, and momentum are fundamental. Energetic causal sets are a working out of this picture.

The energetic causal set models realize the principles and hypotheses of temporal relationalism that I introduced in the previous chapter within a concrete framework. In these principles, time, in the sense of the continual becoming of the present moment, is fundamental to nature. Indeed, our experience of time's passage is the one thing we directly perceive about the world which is truly fundamental. All the rest, including the impression that there are unchanging laws, is approximate and emergent. This view, and the case for it, had been developed during a long collaboration with Roberto Mangabeira Unger. An important consequence is that the laws of nature, rather than being timeless, evolve in time. This reverses the belief, common among physicists, that time is not present in the most fundamental laws, but rather emerges from those laws. Instead, we argue that time, in the sense of the present moment and its passage, is fundamental, while the laws are emergent and subject to change.

Marina Cortês insisted that the laws at the most fundamental level must be irreversible, in two senses. First, the laws are not the same if you reverse the direction of time. If you take a video of a lawful process, you do not get another lawful process by playing it backward. This directly contradicts a widely held belief that the laws of nature are unchanged if you reverse the direction of time.

But all the known fundamental laws, including quantum mechanics, general relativity, and the standard model, are invariant under such a time reversal.* There must be more-fundamental laws, which are not reversible. This raises two challenges: First, can we

* For the experts, a CPT transformation.

invent candidates for an irreversible fundamental law? Second, might it happen that reversible laws emerge as an approximation to more fundamental irreversible laws? These were the questions which energetic causal set models were invented to address.

Cortês also insisted on a deeper sense in which a theory that takes events as primary is irreversible. An event is something that happens. As we stated above, once something happens it cannot *un-happen.* However, the effect of an event can be reversed. If an event changes A to B, it can be followed by an event that changes B back to A. But that makes a history with two events. Once an event has happened it is in the past, and that fact cannot be erased by a future event, even if that future event reverses the effect of the original event.

This thought led us to view the passage of time as a process by which new events are steadily created from present events. While we may give diverse meanings to the word "time," we posited that the passage of time expresses an active process of creation and that this "activity of time" is the creation of novel events, each one after the other.

More specifically, we invented several models, for the purposes of making a concrete realization of our principles and hypotheses. In one model we studied, each event is created from two "parent" events, and then, in turn, becomes the parent of two "child" events.

At each stage in the process there is a vanguard of events, which have been created but have yet to have had all their children. These events make up what we call "the present," as they are the events that will still influence the future.

This process of the continual becoming of events creates a history.

Once an event has had its full allotment of children, it may no longer play a direct role in creating the future, so we say it is in the past. Each past event has a causal past, consisting of those prior events that have directly or indirectly influenced it. Its causal future is the continually growing set of events it directly or indirectly influences. Thus, the past has the structure of a causal set.

We next added energy and momentum, making our model of a growing future an energetic causal set. Each event has a total energy and a total momentum, which are the sums of those of their parents. These are divided up and passed down to their children.

To complete this model, we must answer two questions. How does the process that creates new events out of present events, which we called the activity of time, choose which pair of present events will be the next parents of a novel event? Second, how do events distribute energy and momentum to their children? To answer these questions, we need to prescribe a rule for the creation of new events.

In choosing this rule we were guided by two of the principles I enunciated earlier. The theory should be background independent, which in this context means that the different events should be named, or labeled or distinguished, only by dynamically created structures. Moreover, these structures should not refer to the order in which the events were created. These requirements are satisfied if events are labeled or described only by the structure of their causal pasts.

This makes it natural to invoke the identity of indiscernibles as our second principle. If events are distinguished by their causal pasts, then the causal past of each event must be unique. The event creation rule should then ensure that each event it creates has a causal past different from all the others so far created.

In the models I studied with Cortês, we found two very interesting results. The first, already mentioned, is that the inverse problem appears to be solved, in that there emerges a spacetime into which the events and their causal relations can be mapped. We also found that the systems begin in a very time-asymmetric and disordered phase, which evolves to a phase that is ordered and approximately time symmetric.*

We thus learned an important lesson from the energetic causal set models, which is that time-reversible laws can emerge from more fundamental, irreversible laws. This contradicts the way most physicists think about irreversibility.

WE BEGAN in the last chapter with five principles, which are all ways of expressing Leibniz's principle of sufficient reason, and three hypotheses, which express the fundamental and irreversible character of time and the contrasting, emergent, and contingent nature of space. The theory we seek, which would complete Einstein's twin revolutions, I believe may be the consistent expression of all of these. But, before going all the way there, we introduced several models, which were not meant to be the complete theory, but rather explorations of some aspects it may have by applying only a subset of the principles.

The real ensemble formulation is a relational hidden variable theory. It is not a full application of the principles, as it is situated in a fixed background of time and space, but otherwise it takes the principle of the identity of indiscernibles extremely seriously when it postulates that two events, which have the same view of the universe, are to be identified. I then postulated that the reason two bodies interact more strongly when they are nearer in space is in fact that their views of the rest of the universe are similar. That is, I propose to explain the principle of locality as arising from a deeper principle of similarity of views. To ensure that the identity

* A few years later we understood this two-phased behavior in terms of the dynamics of a general class of deterministic dynamical systems, with a finite number of possible states. Such systems evolve to cycles, and the two phases are the phase of convergence to a cycle followed by cyclic behavior. But a cycle is reversible, because each event has a single child and a single parent.

of indiscernibles is realized, we introduce a force between subsystems that seeks to increase their distinctiveness, or maximize the overall variety. This, as I described earlier, leads to a derivation of quantum mechanics.

Energetic causal sets are models of discrete or quantum universes that explore the hypotheses we made about space and time. In particular, they embody the idea that there is no background space or spacetime. Instead, they take to be fundamental an active, irreversible notion of time and causation, as well as energy and momentum. Spacetime, and space, are emergent and contingent.

The next step is to marry these two models, giving us a relational hidden variable theory that is also background independent, and which realizes the hypothesis that space and locality are emergent.

These two models started as separate research programs, but they shared a motif, which is the central role played by the similarities and differences among events. Both models take these as fundamental, while locality is demoted to an accidental and emergent aspect of nature. It slowly dawned on me that these were different perspectives on a single picture, and so one summer day I sat down and opened a fresh notebook to see if I could tell the full story.

It was immediately clear that the protagonist of this new story is the view. That is, the basic variables are nothing but the views of the universe seen from each event. So I began by fashioning an approach to physics in which these views are fundamental, rather than derived from a more basic structure. In this new perspective, the fundamental laws involve directly only the views and their differences. I call this theory the *causal theory of views*.[6]

The view of an event is nothing but the information available to it from its causal past. The view into the past of an event is like the sky; it is what you see when you look around. Because the speed of light is finite, looking around means looking back, into your past.

The view of an event, as I use the word here, is entirely real and has nothing to do with opinion.* In the theory I am describing, what is real and objective in the world is the information available at each event making up the history of the world, coming to that event from its causal past.

Look up! Your view of the world is like a film projected on a two-dimensional sphere, which we call the sky. The view of an event in a model with three (emergent) dimensions of space will then be represented as a two-dimensional sphere that we call the event's sky. What an event sees on its sky is the events directly in its causal past. More precisely, it sees the energy and momentum coming from each of its parent events. Each parent shows up as a colored point on that event's sky. Each of these points represents a quantum of energy and momentum that has arrived from a past event. The location of each point on the sky records the direction of the momentum, while the color indicates the magnitude of the energy received.

The next step is simple: hypothesize that all that the universe consists of is these skies—each one the view of some event. Rather than construct the views from the causal relations, reverse things and derive the causal relations and everything else from the views. This can work because information contained in the totality of views is enough to reconstruct the causal relations and hence the full history.

As in the real ensemble theory, the laws include the requirement that the variety of all these views is maximized. This has a

* I should warn the reader not to be misled by the colloquial understanding of a "view" in which it stands for the subjective opinion of an individual.

similar effect of leading to the quantum force. Using this, one can derive quantum mechanics as an approximation to the theory.

Here is a one-sentence summary of this theory: the universe consists of nothing but views of itself, each from an event in its history, and the laws act to make these views as diverse as possible.

From here the story unfolds very much like that of the real ensemble theory. Similar views interact with each other, as a result of the mandate to evolve in the direction of ever more diversity. This leads to the emergence of space and of locality in that space. Nonlocality also emerges as interactions which are distant in the emergent space but nearby in terms of similarity of views. Finally, as in the real ensemble formulation, quantum mechanics arises from these nonlocal interactions as an approximate description of the dynamics of views.

The causal theory of views is then a route to a completion of quantum mechanics. It is a realist completion, because it is a theory of beables, which are the views themselves. Most important, it demonstrates that a single fundamental theory can be at the same time a completion of quantum mechanics and an atomic model of spacetime. It can explain the emergence of both locality and nonlocality, of both spacetime and quantum mechanics.

This theory is still only part of the story, and there is still much to learn about it, but it is a way the world might be.

FOR US REALISTS, quantum mechanics cannot be the final story. There is still much to discover. Nonetheless, I remain confident that nature is comprehensible. I am optimistic that the universal power of reasoning that each of us has, together with our vast powers of imagination and our ability to invent novel ideas, will suffice to comprehend the universe. I am especially hopeful about a future in which our individual powers are combined and disciplined by our participation in the scientific community. While I find myself at times deeply frustrated by our lack of definite progress on fundamental physics during this last half century, I am optimistic about the long run. I am confident that in the future our descendants will know vastly more about nature than we do.

I am also sure that the answer to the questions that have bedeviled us for nearly a century will be simple, and expressed in terms of elegant hypotheses and principles of the kind I have put forward here. It would be fortunate indeed if we already have among our library of ideas the answer to how to complete Einstein's twin revolutions. But if we don't, I have no doubt our descendants will, so long as we keep the great adventure of science alive.

EPILOGUE/REVOLUTIONS

Note to Self

> The truth is out there.
>
> —*THE X-FILES*

> Never, never, never, never, never give up.
>
> —DAVID GROSS

Einstein told us that we scientists are opportunists who are willing to break the rules and bend the scientific method to our purpose of discovering how nature works. Each scientist is like an entrepreneur, who has a certain amount of capital to invest; for a theoretical physicist that capital consists mainly of time and attention. The most important decisions we make are what problems we work on and which approaches we choose. Which new paper do we study, to which conferences do we travel, and, once there, to which talks do we listen? The rewards come in different forms: the thrill of discovery, the admiration of one's peers and students, and also one's career, job opportunities, and salary.

If you are interested only in applying the known laws of physics to broaden our appreciation of how nature works, this is a great period to be a physicist. Beautiful discoveries light the way in condensed matter theory, and we are doing real astronomy using gravitational waves to see the universe. These paradigms are working. Steady progress in mathematics drives advances in mathematical physics, with truly brilliant people leading the way to a better understanding of the mathematical structures of our established and nascent theories. Advances in experimental technique are equally impressive, with Moore's law paying off in exponentially increasing range and accuracy of astronomical observations. There is nothing wrong with any of this except that little of it addresses the big foundational puzzles. It is only when we try to advance the project of discovering the fundamental laws and principles that we seem to be spinning our wheels.

At the present moment in fundamental physics and cosmology, there are basically only two ways to bet. We either bet that we know all the fundamental principles, or we bet there are basic ideas and principles missing. The major research programs, such as inflation, string theory, and loop quantum gravity, are all ways of betting that we know the basic principles of fundamental physics. With notable exceptions, workers in these fields take for granted that the basic principles of quantum theory and relativity are sound and apply to the new theory. Many of the people who work outside these programs are doing so because they bet there is much more to be discovered. People like me, who do some of both, are hedging our bets.

When it comes to quantum mechanics we face the same choice. Either we bet that we have the complete theory in our hands and just need to understand it better, or we bet the theory is incomplete in important ways. The Copenhagen interpretation, the operational interpretations, Everett quantum mechanics, and so on are all ways of betting we know everything important about quantum phenomena. Anyone who focuses exclusively on one of the realist proposals such as pilot wave theory or spontaneous collapse is betting their favorite theory will turn out to be the correct completion of quantum mechanics. In either case, the bet assumes that we know all the principles needed to understand nature.

What about those of us who are convinced that a completion is needed, but are not convinced any of the well-studied ones have the ring of truth? How are we to bet?

Up to now, my own bets have fallen on both sides of these divides. My most successful bets employed ideas and technical tools from particle physics to solve problems in quantum gravity. This was one of the routes that led to loop quantum gravity. But from time to time I wrote papers reporting my efforts to invent relational hidden variable theories. And the very best of my early papers was an attempt to connect the principle of inertia to quantum foundations. As the years went on I extended my foundational efforts to the landscape issue, which led to my work on the nature of time. But my bread-and-butter work remains quantum gravity, both the phenomenology of the theory and loop quantum gravity.

A book project is a kind of mental therapy, which forces you to examine your confused thoughts and intuitions and develop them to their logical conclusions. So now that I have written a book which argues that a radically new theory is needed to solve the foundational issues in physics and cosmology, what am I going to do about it? Do I keep to the same safe, hedged program, or go all out on an attempt to solve the real problems?

To bet that the truth requires something as yet undiscovered, we must spend our time searching for that unknown completion. We can't just sail down one shoreline and up another. We head west: out of sight of land, following our own compass, or the best facsimile thereof that we can cobble together from the clues we take seriously.

There is no more reasonable bet than that our current knowledge is incomplete. In every era of the past our knowledge was incomplete; why should our period be any different? Certainly the puzzles we face are at least as formidable as any in the past. But almost no one bets this way. This puzzles me.

I suspect it's hard for many physicists to imagine that we are not near the end of our search for the ultimate laws of nature. We have been raised in a culture in which it's all about having the right answer, and we owe our careers to having been the scientists who had them. But I've always had in my head an image of how much more people in the future will know, and how silly our claims to knowledge will look to them. This has probably made me a less effective advocate of my own ideas.

So what do we do with the partly successful inventions, such as loop quantum gravity? At first, the discovery of a new possible direction, incomplete and without experimental confirmation (in other words, highly vulnerable to criticism, as most new theories are at birth), is very worth our time and focus. That X, however incompletely formulated, is something that just might be true, or be part of the truth, even without positive evidence, is certainly good for a decade of examination. But after a third or more of a century, during which many career-long efforts have failed to budge *might be true* any closer to *must be true*, isn't it time to move on? You might think I'm repeating polemics from the string wars, but I'm thinking, with a great deal of affection, of all of us whose years of hard work have failed to yield the breakthroughs we fantasized about. Including myself; especially myself.

Why do we write more and more papers on approaches whose deficiencies have been obvious for decades, and almost no papers proposing new completions of quantum mechanics? It is not for lack of caring, for everyone I know who works on quantum foundations has chosen that risky path because they care passionately about how nature resolves the measurement problem and the other puzzles.

I, for one, am tired of arguing over the ins and outs and relative merits of the existing approaches, and the clever fixes invented to save an idea that is pretty obviously collapsing from insufficiency. So I have a decision to make: I either keep on the present path, which will end up on the top of that low hill just past the next village, or head down into the swamps to stumble along unknown paths in search of undiscovered mountains. If I take the swamp trail, I will almost certainly fail, but I hope to send back reports to interest and inspire those few others who feel in their bones the cost of our ignorance, of giving up the search too soon.

Even if I'm convinced that something very new is needed, I have little idea how to search for scientific truth except by building on an existing research program, using a well-honed tool kit and methodology. This is research as it is taught, recognized, funded, and rewarded by the academic community. A community, I should mention, that it is necessary to be an active part of to get your work taken seriously by people who know enough to evaluate it. What would I put in my research proposals, if my ideas are not expressible in the language of an already existing and widely followed research program? What problems do I set for my PhD students, if they are not to calculate something using tools developed within a given framework? Do I tell my students to wake up in the morning, make coffee, open a blank notebook, and stare at it until a disheveled angel arrives with a revelation? Is that what I should do myself? How many days, weeks, months, years, how many incoherent scribbled pages, do I tolerate before giving up?

It is not just that to try to invent a whole new physics is risky for my career and damaging to my emotional stability. I don't even know how to begin. Almost no one alive has done that, in the way that the revolutionaries of a century ago did. In my experience there is little as terrifying as putting aside the basic principles that form the foundation of our understanding of how we fit into nature—isn't that why it feels comforting to know them?

It certainly is easier to work within an existing framework, to test the limits of what we know from the inside, so to speak. We can do this while keeping an open mind about the basic principles and looking out for opportunities to modify those principles or even introduce new ones. Even more important is to keep on the lookout for new opportunities to test theories against experiments and observations. This is what I have done for most of my career, and I venture to say this is true as well of many who work on the main approaches, such as string theory and loop quantum gravity. What we have to show for this is a collection of beautiful results, which may or may not lead to the true story, and, especially precious, a few proposals for new principles, including the holographic principle and the principle of relative locality. But, with all due respect to those of us who invested most of our time in reasonable approaches to the development of reasonable theories, it does not seem to have been enough, this time.

I say to myself, I'll take such risks after I get my PhD, after I get my postdoc, after my faculty position, after tenure. But even tenured, senior, famous professors must apply for research grants, and there is always that fancy career-culminating prize, or that comfortable and prestigious chair. So we'll just wait till retirement. Then we'll be free to take the big risks. Well, as someone closing in on that, I can report that the only thing you learn is certain, as your fifties and sixties rush by, every day busy with a schedule full of seminars, faculty meetings, working with students, classes, review panels, airplanes, hotels, and conference talks, is that you are not immortal.

So maybe it's all up to a brilliant student somewhere, impossibly arrogant, as the young Einstein was, but blindingly talented enough to absorb the essentials of all we have done, before putting them to one side and confidently starting over.

A friend once told me that the academic world was modeled on monasteries, which were designed to perpetuate old knowledge while resisting the new. Even after decades in the system I am amazed at how the fine mechanics of this work. There is no arguing with the logic of academic fame, which rewards every scientific success with distractions that make it harder to do more science, while imposing enormous disincentives to putting aside polishing your legacy to take on new challenges.

The academic world is very well suited to support what Thomas Kuhn called normal science. That is great until it becomes long past due to complete a revolution.

To my knowledge, few have stumbled on a major discovery by accident; most true breakthroughs were found after years and years of hard, unrewarding work. Feynman said to discover something new you have to take the time to make every mistake possible along the way. And he surely knew.

So I have no better answer than to face the blank notebook. We do have role models. Einstein did it. Bohr did it. De Broglie, Schrödinger, and Heisenberg did it, as did Bohm and Bell. They each found a path from that blank page to a foundational discovery that enlarged our understanding of how nature works. Start by writing down what you are confident we know for sure. Ask yourself which of the fundamental principles of the present canon must survive the coming revolution. That's the first page. Then turn again to a blank page and start thinking.

ACKNOWLEDGMENTS

This book represents a lifetime of wrestling with the puzzles of quantum foundations, and I have to thank, first of all, Herbert Bernstein, for his revolutionary freshman quantum mechanics course, for making me the grader in the course to make sure I learned how to solve the problems, and for many years of friendship since then. In graduate school I was fortunate to be able to study with Abner Shimony, who has been a role model for all of us wishing to bring the rigor and depth of philosophy to the examination of foundational problems in physics. For that I have to thank Hilary Putnam, who told me that Abner would be able to answer my questions about quantum theory that he wasn't able to.

In graduate school and in the years since, I have been fortunate to meet and converse with some of the truly deep thinkers who inspire us still: Steve Adler, Yakir Aharonov, Bryce DeWitt, Cécile DeWitt-Morette, Freeman Dyson, Paul Feyerabend, Richard Feynman, Jim Hartle, Gerard 't Hooft, Chris Isham, Edward Nelson, Roger Penrose, Leonard Susskind, John Archibald Wheeler, and Eugene Wigner.

Shortly after receiving my PhD, I met Julian Barbour, who introduced me to Leibniz and Mach, and has been my mentor and guide to relational philosophy since. My philosophical education has continued through conversations with David Albert, Harvey Brown, Jim Brown, Jeremy Butterfield, Jenann Ismael, and Steve Weinstein, among many. Henrique Gomes, Simon Saunders, Roderich Tumulka, Antony Valentini, and David Wallace have been especially helpful reading and commenting on drafts and patiently explaining what I got wrong. All errors that remain are, however, my responsibility.

Then I want very much to thank those who have become friends through our shared work on foundational problems: Stephon Alexander, Giovanni Amelino-Camelia, Abhay Ashtekar, Eli Cohen, Marina Cortês, Louis Crane, John Dell, Avshalom Elitzur, Laurent Freidel, Sabine Hossenfelder, Ted Jacobson, Stuart Kauffman, Jurek Kowalski-Glikman, Andrew Liddle, Renate Loll, João Magueijo, Roberto Mangabeira Unger, Fotini Markopoulou, and Carlo Rovelli.

The book has been very much improved by feedback from Krista Blake, Saint Clair Cemin, Dina Graser, Jaron Lanier, and Donna Moylan. I also want to thank Kaća Bradonjić for the illustrations and for many wise and helpful suggestions on the text.

For helpful conversations and correspondence on specific points, I must thank Jim Baggott, Julian Barbour, Freeman Dyson, Olival Freire, Stuart Kauffman, Michael Nielsen, Philip Pearle, Bill Poirier, Carlo Rovelli, and John Stachel. Alexander Blum and Jürgen Renn helped me tell a true story of the history of quantum mechanics.

I am extremely grateful to be part of a vibrant community at the Perimeter Institute for Theoretical Physics focused on foundational physics, which gives me a home and a context for my work. In addition to those already named, I've learned immeasurably over the years from Gemma De las Cuevas, Bianca Dittrich, Fay Dowker, Chris Fuchs, Lucien Hardy, Adrian Kent, Rafael Sorkin, Rob Spekkens, and many others. I wish to thank Mike Lazaridis, Howard

Burton, and Neil Turok for including me in this adventure of a lifetime, and also give a shout-out to Michael Duschenes and the whole PI staff for their intelligent and dedicated work.

I am grateful to several classes of students, going back to "Nature Loves to Hide" at Hampshire College, who have taken various versions of a quantum physics for poets class, during which I tested the pedagogical strategies I use here. Most recently, Camilla Singh allowed herself to be a test case in an experiment to teach quantum mechanics to artists.

John Brockman, Katinka Matson, and Max Brockman have been my literary agents and friends for the many years I have been writing books. Scott Moyers, Christopher Richards, and Thomas Penn have been great editors, and I am especially grateful to them for insisting that I could write a better book than I knew. I am proud to be among the many writers who have benefited from the critical eye of Louise Dennys.

Finally, I owe everything to Dina Graser and Kai Smolin, who have supported me throughout all the ups and downs of this project.

NOTES

Preface

1. J. S. Bell, "On the Einstein Podolsky Rosen Paradox," *Physics* 1, no. 3 (November 1964): 195–200.

Chapter 1: Nature Loves to Hide

Epigraph Albert Einstein, "A Reply to Criticisms," *Albert Einstein: Philosopher-Scientist*, ed. P. A. Schillp, 3rd ed. (Peru, IL: Open Court Publishing, 1988).

1. Einstein to Max Born, December 4, 1926, in *The Born-Einstein Letters: The Correspondence Between Albert Einstein and Max and Hedwig Born, 1916–1955, with Commentaries by Max Born*, trans. Irene Born (New York: Walker and Co., 1971) 88.

Chapter 2: Quanta

1. Tom Stoppard, *Arcadia: A Play*, first performance, Royal National Theatre, London, April 13, 1993; act 1, scene 1 (New York: Farrar, Straus and Giroux, 2008), 9.

Chapter 4: How Quanta Share

Epigraph John Archibald Wheeler, *Quantum Theory and Measurement*, ed. J. A. Wheeler and W. H. Zurek (Princeton: Princeton University Press, 1983): 194.

1. Albert Einstein, Boris Podolsky, and Nathan Rosen, "Can Quantum-Mechanical Description of Physical Reality Be Considered Complete?," *Physical Review* 47, no. 10 (May 15, 1935): 777–80.
2. Alain Aspect, Philippe Grangier, and Gérard Roger, "Experimental Tests of Realistic Local Theories via Bell's Theorem," *Physical Review Letters* 47, no. 7 (August 1981): 460–63; Alain Aspect, Jean Dalibard, and Gérard Roger, "Experimental Test of Bell's Inequalities Using Time-Varying Analyzers," *Physical Review Letters* 49, no. 25 (December 1982): 1804–7.
3. Niels Bohr, "Can Quantum-Mechanical Description of Physical Reality Be Considered Complete?," *Physical Review* 48, no. 8 (October 1935): 696–702.
4. Simon Kochen and E. P. Specker, "The Problem of Hidden Variables in Quantum Mechanics," *Journal of Mathematics and Mechanics* 17, no. 1 (July 1967): 59–87; John S. Bell, "On the Problem of Hidden Variables in Quantum Mechanics," *Reviews of Modern Physics* 38, no. 3 (July 1966): 447–52.

Chapter 6: The Triumph of Anti-Realism

Epigraph Christopher A. Fuchs and Asher Peres, "Quantum Theory Needs No 'Interpretation,'" *Physics Today* 53, no. 3 (March 2000): 70–71, https://doi.org/10.1063/1.883004.
1. J. J. O'Connor and E. F. Robertson, "Louis Victor Pierre Raymond duc de Broglie," http://www-history.mcs.st-andrews.ac.uk/Biographies/Broglie.html.
2. Louis de Broglie, interview by Thomas S. Kuhn, Andre George, and Theo Kahan, January 7, 1963, transcript, Niels Bohr Library & Archives, American Institute of Physics, College Park, MD, https://repository.aip.org/islandora/object/nbla:272502.
3. Werner Heisenberg, *The Physicist's Conception of Nature*, trans. Arnold J. Pomerans (New York: Harcourt Brace, 1958), 15, 29.
4. Niels Bohr (1934), quoted in Max Jammer, *The Philosophy of Quantum Mechanics: The Interpretations of Quantum Mechanics in Historical Perspective* (New York: John Wiley and Sons, 1974), 102.

Chapter 7: The Challenge of Realism: de Broglie and Einstein

1. Guido Bacciagaluppi and Antony Valentini, *Quantum Theory at the Crossroads: Reconsidering the 1927 Solvay Conference* (Cambridge, UK: Cambridge University Press, 2009), 235.
2. Bacciagaluppi and Valentini, 487.

3. Grete Hermann, "*Die naturphilosophischen Grundlagen der Quantenmechanik*," *Die Naturwissenschaften* 23, no. 42 (October 1935), 718–21, doi:10.1007/BF01491142; Grete Hermann, "The Foundations of Quantum Mechanics in the Philosophy of Nature," trans. with an introduction by Dirk Lumma, *The Harvard Review of Philosophy* 7, no. 1 (1999): 35–44.
4. John Bell, "Interview: John Bell," interview by Charles Mann and Robert Crease, *Omni* 10, no. 8 (May 1988): 88.
5. N. David Mermin, "Hidden Variables and the Two Theorems of John Bell," *Reviews of Modern Physics* 65, no. 3 (July 1993): 805–6.

Chapter 8: Bohm: Realism Tries Again

Epigraph Roderich Tumulka, "On Bohmian Mechanics, Particle Creation, and Relativistic Space-Time: Happy 100th Birthday, David Bohm!," *Entropy* 20, no. 6 (June 2018): 462, arXiv:1804.08853v3.

1. David Bohm, "A Suggested Interpretation of Quantum Theory in Terms of 'Hidden' Variables, 1," *Physical Review* 85, no. 2 (January 1952): 166–79.
2. Albert Einstein, quoted in Wayne Myrvold, "On Some Early Objections to Bohm's Theory," *International Studies in the Philosophy of Science* 17, no. 1 (March 2003): 7–24.
3. Albert Einstein, quoted in E. David Peat, *Infinite Potential: The Life and Times of David Bohm* (New York: Basic Books, 1997), 132.
4. Albert Einstein, "*Elementäre Überlegungen zur Interpretation der Grundlagen der Quanten-Mechanik*," in *Scientific Papers Presented to Max Born* (New York: Hafner, 1953), 33–40; quoted in Myrvold.
5. Benyamin Cohen, "4 Things Einstein Said to Cheer Up His Sad Friend," From the Grapevine, June 13, 2017, https://www.fromthegrapevine.com/lifestyle/einstein-bohm-letters-winner-auction-israel.
6. Werner Heisenberg, quoted in Myrvold, "On Some Early Objections," 12.
7. Olival Freire Jr., "Science and Exile: David Bohm, the Hot Times of the Cold War, and His Struggle for a New Interpretation of Quantum Mechanics," *Historical Studies on the Physical and Biological Sciences* 36, no. 1 (September 2005): 1–34, https://arxiv.org/pdf/physics/0508184.pdf.
8. J. Robert Oppenheimer remarks to Max Dresden, in Max Dresden, *H. A. Kramers: Between Tradition and Revolution* (New York: Springer-Verlag, 1987), 133. Also quoted in F. David Peat's *Infinite Potential: The Life and Times of David Bohm* (Reading, MA: Addison-Wesley, 1996), where he attributes them to Dresden's "remarks from the floor at the American Physical Society Meeting, Washington, May, 1989. Dresden confirmed

this version in an interview with the author [Peat] immediately following that session and in a letter to the author." (Quote, p. 133; note, p. 334.)

9. Peat, *Infinite Potential*, 133.
10. John Nash to J. Robert Oppenheimer, July 10, 1957, Institute for Advanced Study, Shelby White and Leon Levy Archives Center, https://www.ias.edu/ideas/2015/john-forbes-nash-jr.
11. Léon Rosenfeld to David Bohm, May 30, 1952, quoted in Louisa Gilder, *The Age of Entanglement: When Quantum Physics Was Reborn* (New York: Alfred A. Knopf, 2008), 216–17.
12. Antony Valentini, "Signal-Locality, Uncertainty, and the Sub-Quantum H-Theorem, 1," *Physics Letters A* 156, nos. 1–2 (June 1991): 5–11; "2," *Physics Letters A* 158, nos. 1–2 (August 1991): 1–8.
13. Antony Valentini and Hans Westman, "Dynamical Origin of Quantum Probabilities," *Proceedings of the Royal Society of London A* 461, no. 2053 (January 2005): 253–72, arXiv:quant-ph/0403034; Eitan Abraham, Samuel Colin, and Antony Valentini, "Long-Time Relaxation in Pilot-Wave Theory," *Journal of Physics A: Mathematical and Theoretical* 47, no. 39 (September 2014): 5306, arXiv:1310.1899.
14. Antony Valentini, "Signal-Locality in Hidden-Variables Theories," *Physics Letters A* 297, nos. 5–6 (May 2002): 273–78.
15. Nicolas G. Underwood and Antony Valentini, "Anomalous Spectral Lines and Relic Quantum Nonequilibrium" (2016), arXiv:1609.04576; Samuel Colin and Antony Valentini, "Robust Predictions for the Large-Scale Cosmological Power Deficit from Primordial Quantum Nonequilibrium," *International Journal of Modern Physics* D25, no. 6 (April 2016): 1650068, arXiv:1510.03508.

Chapter 9: The Collapse of the Quantum State

1. David Bohm and Jeffrey Bub, "A Proposed Solution of the Measurement Problem in Quantum Mechanics by a Hidden Variable Theory," *Reviews of Modern Physics* 38, no. 3 (July 1966): 453–69.
2. Philip Pearle, "Reduction of the State Vector by a Nonlinear Schrödinger Equation," *Physical Review D* 13, no. 4 (February 1976): 857–68.
3. Giancarlo Ghirardi, Alberto Rimini, and Tullio Weber, "Unified Dynamics for Microscopic and Macroscopic Systems," *Physical Review D* 34, no. 2 (July 1986): 470–91.
4. Roderich Tumulka, "A Relativistic Version of the Ghirardi-Rimini-Weber Model," *Journal of Statistical Physics* 125, no. 4 (November 2006): 821–40.

5. Roger Penrose, "Gravitational Collapse and Space-Time Singularities," *Physical Review Letters* 14, no. 3 (January 1965): 57–59.
6. Stephen W. Hawking and Roger Penrose, "The Singularities of Gravitational Collapse and Cosmology," *Proceedings of the Royal Society A* 314, no. 1519 (January 1970): 529–48.
7. R. Penrose, "Time-Asymmetry and Quantum Gravity," in *Quantum Gravity 2: A Second Oxford Symposium*, eds. C. J. Isham, R. Penrose, and D. W. Sciama (Oxford: Clarendon Press, 1981), 244; R. Penrose, "Gravity and State Vector Reduction," in *Quantum Concepts in Space and Time*, eds. R. Penrose and C. J. Isham (Oxford: Clarendon Press, 1986), 129; R. Penrose, "Non-locality and Objectivity in Quantum State Reduction," in *Quantum Coherence and Reality: In Celebration of the 60th Birthday of Yakir Aharonov*, eds. J. S. Anandan and J. L. Safko (Singapore: World Scientific, 1995), 238; R. Penrose, *Shadows of the Mind: A Search for the Missing Science of Consciousness* (Oxford: Oxford University Press, 1994); R. Penrose, "On Gravity's Role in Quantum State Reduction," *General Relativity and Gravitation* 28, no. 5 (May 1996): 581–600; I. Fuentes and R. Penrose, "Quantum State Reduction via Gravity, and Possible Tests Using Bose-Einstein Condensates," in *Collapse of the Wave Function: Models, Ontology, Origin, and Implications*, ed. S. Gao (Cambridge, UK: Cambridge University Press, 2018), 187.
8. L. Diósi, "Models for Universal Reduction of Macroscopic Quantum Fluctuations," *Physical Review A* 40, no. 3 (August 1989): 1165–74; F. Károlyházy, "Gravitation and Quantum Mechanics of Macroscopic Bodies," *Il Nuovo Cimento A* 42, no. 2 (March 1966): 390–402; F. Károlyházy, A. Frenkel, and B. Lukács, "On the Possible Role of Gravity in the Reduction of the Wave Function," in *Quantum Concepts in Space and Time*, 109–28.
9. S. Bose, A. Mazumdar, G. W. Morley, H. Ulbricht, M. Toros, M. Paternostro, A. A. Geraci, P. F. Barker, M. S. Kim, and G. Milburn, "Spin Entanglement Witness for Quantum Gravity," *Physical Review Letters* 119, no. 24 (December 2017): 240401, arXiv:1707.06050; C. Marletto and V. Vedral, "Gravitationally Induced Entanglement between Two Massive Particles Is Sufficient Evidence of Quantum Effects in Gravity," *Physical Review Letters* 119, no. 24 (December 2017): 240402, arXiv:1804.11315.
10. Philip Pearle, "A Relativistic Dynamical Collapse Model," *Physical Review D* 91, no. 10 (May 2015): 105012, arXiv:1412.6723.
11. Rodolfo Gambini and Jorge Pullin, "The Montevideo Interpretation of Quantum Mechanics: A Short Review," *Entropy* 20, no. 6 (February 2015): 413, arXiv:1502.03410.
12. Stephen L. Adler, "Gravitation and the Noise Needed in Objective Reduction Modes," in *Quantum Nonlocality and Reality: 50 Years of Bell's*

Theorem, eds. Mary Bell and Shan Gao (Cambridge, UK: Cambridge University Press, 2016), 390–99.

Chapter 10: Magical Realism

Epigraph Bryce S. DeWitt, "Quantum Mechanics and Reality: Could the Solution to the Dilemma of Indeterminism Be a Universe in Which All Possible Outcomes of an Experiment Actually Occur?" *Physics Today* 23, no. 9 (September 1970): 155–65.

1. Hugh Everett III, "'Relative State' Formulation of Quantum Mechanics," *Reviews of Modern Physics* 29, no. 3 (July 1957): 454–62.

Chapter 11: Critical Realism

1. David Deutsch, "Quantum Theory of Probability and Decisions," *Proceedings of the Royal Society A* 455, no. 1988 (August 1999): 3129–37, arXiv:quant-ph/9906015.
2. David Wallace, "Quantum Probability and Decision Theory, Revisited" (2002), arXiv:quant-ph/0211104; Wallace, "Everettian Rationality: Defending Deutsch's Approach to Probability in the Everett Interpretation," *Studies in History and Philosophy of Science Part B: Studies in History and Philosophy of Modern Physics* 34, no. 3 (September 2003): 415–39, arXiv:quant-ph/0303050; Wallace, "Quantum Probability from Subjective Likelihood: Improving on Deutsch's Proof of the Probability Rule," *Studies in History and Philosophy of Science Part B: Studies in History and Philosophy of Modern Physics* 38, no. 2 (June 2007): 311–32, arXiv:quant-ph/0312157; Wallace, "A Formal Proof of the Born Rule from Decision-Theoretic Assumptions" (2009), arXiv:quant-ph/0906.2718; Simon Saunders, "Derivation of the Born Rule from Operational Assumptions," *Proceedings of the Royal Society A* 460, no. 2046 (June 2004): 1771–88, arXiv:quant-ph/0211138.
3. Lawrence S. Schulman, "Note on the Quantum Recurrence Theorem," *Physical Review A* 18, no. 5 (November 1978): 2379–80, doi:10.1103/PhysRevA.18.2379.
4. Steven Weinberg, "The Trouble with Quantum Mechanics," *The New York Review of Books*, January 19, 2017, https://www.nybooks.com/articles/2017/01/19/trouble-with-quantum-mechanics/.

Chapter 12: Alternatives to Revolution

Epigraph Lucien Hardy, "Reformulating and Reconstructing Quantum Theory" (2011), arXiv:1104.2066.

1. Richard Feynman, "Simulating Physics with Computers," keynote address delivered at the MIT Physics of Computation Conference, 1981. Published in *International Journal of Theoretical Physics* 21, nos. 6–7 (June 1982): 467–88.
2. David Deutsch, "Quantum Theory, the Church-Turing Principle and the Universal Quantum Computer," *Proceedings of the Royal Society A* 400, no. 1818 (July 1985): 97–117.
3. John Archibald Wheeler, "Information, Physics, Quantum: The Search for Links," in *Proceedings of the 3rd International Symposium: Foundations of Quantum Mechanics in the Light of New Technology, Tokyo, 1989*, eds. Shunichi Kobayashi et al. (Tokyo: Physical Society of Japan, 1990), 354–58.
4. John Archibald Wheeler, quoted in Paul Davies, *The Goldilocks Enigma*, also titled *Cosmic Jackpot* (Boston and New York: Houghton Mifflin, 2006), 281.
5. Christopher A. Fuchs and Blake C. Stacey, "QBism: Quantum Theory as a Hero's Handbook" (2016), arXiv:1612.07308.
6. Louis Crane, "Clock and Category: Is Quantum Gravity Algebraic?," *Journal of Mathematical Physics* 36, no. 11 (May 1995): 6180–93, arXiv:gr-qc /9504038; Carlo Rovelli, "Relational Quantum Mechanics," *International Journal of Theoretical Physics* 35, no. 8 (August 1996): 1637–78, arXiv:quant-ph/9609002; Lee Smolin, "The Bekenstein Bound, Topological Quantum Field Theory and Pluralistic Quantum Cosmology" (1995), arXiv:gr-qc/9508064.
7. Ruth E. Kastner, Stuart Kauffman, and Michael Epperson, "Taking Heisenberg's Potentia Seriously" (2017), arXiv:1709.03595.
8. Julian Barbour, *The End of Time: The Next Revolution in Physics* (Oxford: Oxford University Press, 1999).
9. Henrique de A. Gomes, "Back to Parmenides" (2016, 2018), arXiv :1603.01574.

Chapter 13: Lessons

1. I am grateful to Avshalom Elitzur and Eli Cohen for many discussions on these kinds of cases.

2. For a recent review, see Roderich Tumulka, "Bohmian Mechanics," in *The Routledge Companion to the Philosophy of Physics*, eds. Eleanor Knox and Alastair Wilson (New York: Routledge, 2018), arXiv:/1704.08017.
3. Yakir Aharonov and Lev Vaidman, "The Two-State Vector Formalism of Quantum Mechanics: An Updated Review," in *Time in Quantum Mechanics*, vol. 1, eds. J. Gonzalo Muga, Rafael Sala Mayato, and Íñigno Egusquiza, 2nd ed., Lecture Notes in Physics 734 (Berlin and Heidelberg: Springer, 2008), 399–447, arXiv:quant-ph/0105101v2.
4. John G. Cramer, "The Transactional Interpretation of Quantum Mechanics," *Reviews of Modern Physics* 58, no. 3 (July 1986), 647–87; Cramer, *The Quantum Handshake: Entanglement, Nonlocality and Transactions* (Cham, Switzerland: Springer International, 2016); Ruth E. Kastner, "The Possibilist Transactional Interpretation and Relativity," *Foundations of Physics* 42, no. 8 (August 2012): 1094–113.
5. Huw Price, "Does Time-Symmetry Imply Retrocausality? How the Quantum World Says 'Maybe,'" *Studies in History and Philosophy of Science Part B: Studies in History and Philosophy of Modern Physics* 43, no. 2 (May 2012), 75–83, arXiv:1002.0906.
6. Rafael D. Sorkin, "Quantum Measure Theory and Its Interpretation," in *Quantum Classical Correspondence: Proceedings of the 4th Drexel Symposium on Quantum Nonintegrability, Drexel University, Philadelphia, USA, September 8–11, 1994*, eds. Bei-Lok Hu and Da Hsuan Feng (Cambridge, MA: International Press, 1997), 229–51, arXiv:gr-qc/9507057.
7. Murray Gell-Mann and James B. Hartle, "Quantum Mechanics in the Light of Quantum Cosmology," in *Proceedings of the 3rd International Symposium: Foundations of Quantum Mechanics in the Light of New Technology, Tokyo, 1989*, 321–43; Gell-Mann and Hartle, "Alternative Decohering Histories in Quantum Mechanics," in *Proceedings of the 25th International Conference on High Energy Physics, 2–8 August 1990, Singapore*, eds. K. K. Phua and Y. Yamaguchi, vol. 1, 1303–10 (Singapore and Tokyo: South East Asia Theoretical Physics Association and Physical Society of Japan, dist. World Scientific, 1990); Gell-Mann and Hartle, "Time Symmetry and Asymmetry in Quantum Mechanics and Quantum Cosmology," in *Proceedings of the NATO Workshop on the Physical Origins of Time Asymmetry, Mazagón, Spain, September 30–October 4, 1991*, eds. J. Halliwell, J. Pérez-Mercader, and W. Zurek (Cambridge, UK: Cambridge University Press, 1992), arXiv:gr-qc/9304023; Gell-Mann and Hartle, "Classical Equations for Quantum Systems," *Physical Review D* 47, no. 8 (April 1993): 3345–82, arXiv:gr-qc/9210010.
8. Robert B. Griffiths, "Consistent Histories and the Interpretation of Quantum Mechanics," *Journal of Statistical Physics* 36, nos. 1–2 (July 1984),

219–72; Griffiths, "The Consistency of Consistent Histories: A Reply to d'Espagnat," *Foundations of Physics* 23, no. 12 (December 1993): 1601–10; Roland Omnès, "Logical Reformulation of Quantum Mechanics, 1: Foundations," *Journal of Statistical Physics* 53, nos. 3–4 (November 1988): 893–932; Omnès, "Logical Reformulation of Quantum Mechanics, 2: Interferences and the Einstein-Podolsky-Rosen Experiment," ibid., 933–55; Omnès, "Logical Reformulation of Quantum Mechanics, 3: Classical Limit and Irreversibility," ibid., 957–75; Omnès, "Logical Reformulation of Quantum Mechanics, 4: Projectors in Semiclassical Physics," *Journal of Statistical Physics* 57, nos. 1–2 (October 1989): 357–82; Omnès, "Consistent Interpretations of Quantum Mechanics," *Reviews of Modern Physics* 64, no. 2 (April 1992): 339–82.

9. Fay Dowker and Adrian Kent, "On the Consistent Histories Approach to Quantum Mechanics," *Journal of Statistical Physics* 82, nos. 5–6 (March 1996): 1575–646, arXiv:gr-qc/9412067.
10. Michael J. W. Hall, Dirk-André Deckert, and Howard M. Wiseman, "Quantum Phenomena Modeled by Interactions between Many Classical Worlds," *Physical Review X* 4, no. 4 (October 2014): 041013, arXiv:1402.6144.
11. Benhui Yang, Wenwu Chen, and Bill Poirier, "Rovibrational Bound States of Neon Trimer: Quantum Dynamical Calculation of All Eigenstate Energy Levels and Wavefunctions," *Journal of Chemical Physics* 135, no. 9 (September 2011): 094306; Gérard Parlant, Yong-Cheng Ou, Kisam Park, and Bill Poirier, "Classical-like Trajectory Simulations for Accurate Computation of Quantum Reactive Scattering Probabilities," invited contribution and lead article, special issue to honor Jean-Claude Rayez, *Computational and Theoretical Chemistry* 990 (June 2012): 3–17.
12. Gerard 't Hooft, "Time, the Arrow of Time, and Quantum Mechanics" (2018), arXiv:1804.01383.
13. Lee Smolin, "Could Quantum Mechanics Be an Approximation to Another Theory?" (2006), arXiv:quant-ph/0609109.
14. Matthew F. Pusey, Jonathan Barrett, and Terry Rudolph, "On the Reality of the Quantum State," *Nature Physics* 8, no. 6 (June 2012): 475–78, arXiv:1111.3328.

Chapter 14: First, Principles

1. Lee Smolin, *Time Reborn: From the Crisis in Physics to the Future of the Universe* (New York: Houghton Mifflin, 2013); Roberto Mangabeira Unger and Lee Smolin, *The Singular Universe and the Reality of Time: A*

Proposal in Natural Philosophy (Cambridge, UK: Cambridge University Press, 2015); Smolin, "Temporal Naturalism," invited contribution to special issue on Cosmology and Time, *Studies in History and Philosophy of Science Part B: Studies in History and Philosophy of Modern Physics* 52, no. 1 (November 2015): 86–102, arXiv:1310.8539.

2. Fotini Markopoulou and Lee Smolin, "Disordered Locality in Loop Quantum Gravity States," *Classical and Quantum Gravity* 24, no. 15 (July 2007): 3813–24, arXiv:gr-qc/0702044.
3. Lee Smolin, "Derivation of Quantum Mechanics from a Deterministic Non-Local Hidden Variable Theory, I. The Two-Dimensional Theory," IAS preprint PRINT-83-0802 (Princeton: Institute for Advanced Study, August 1983); Smolin, "Stochastic Mechanics, Hidden Variables and Gravity," in *Quantum Concepts in Space and Time*, eds. Roger Penrose and C. J. Isham (Oxford and New York: Clarendon Press / Oxford University Press, 1986).
4. Lee Smolin, "Matrix Models as Non-Local Hidden Variables Theories," in *Quo Vadis Quantum Mechanics?*, eds. Avshalom C. Elitzur, Shahar Dolev, and Nancy Kolenda, The Frontiers Collection (Berlin and Heidelberg: Springer, 2005), 121–52; Smolin, "Non-Local Beables," *International Journal of Quantum Foundations* 1, no. 2 (April 2015): 100–106, arXiv:1507.08576.
5. Stephen L. Adler, *Quantum Theory as an Emergent Phenomenon: The Statistical Mechanics of Matrix Models as the Precursor of Quantum Field Theory* (Cambridge, UK: Cambridge University Press, 2004); book draft, *Statistical Dynamics of Global Unitary Invariant Matrix Models as Pre-Quantum Mechanics* (2002), arXiv:hep-th/0206120.
6. Artem Starodubtsev, "A Note on Quantization of Matrix Models," *Nuclear Physics B* 674, no. 3 (December 2003): 533–52, arXiv:hep-th/0206097.
7. Markopoulou and Smolin, "Disordered Locality."
8. Fotini Markopoulou and Lee Smolin, "Quantum Theory from Quantum Gravity," *Physical Review D* 70, no. 12 (December 2004): 124029, arXiv:gr-qc/0311059.
9. Gottfried Wilhelm Leibniz, *The Monadology*, 1714, in *Leibniz: Philosophical Writings*, ed. G. H. R. Parkinson, trans. Mary Morris and G. H. R. Parkinson (London: J. M. Dent, 1973).
10. Julian Barbour and Lee Smolin, "Extremal Variety as the Foundation of a Cosmological Quantum Theory" (1992), arXiv:hep-th/9203041.
11. Leibniz, *The Monadology*, paragraph 57, in *Leibniz, Philosophical Writings*.
12. Lee Smolin, "The Dynamics of Difference," *Foundations of Physics* 48, no. 2 (February 2018): 121–34, arXiv:1712.04799; Smolin, "Quantum Mechanics and the Principle of Maximal Variety," *Foundations of Physics* 46,

no. 6 (June 2016): 736–58, arXiv:1506.02938; Smolin, "A Real Ensemble Interpretation of Quantum Mechanics," *Foundations of Physics* 42, no. 10 (October 2012): 1239–61, arXiv:1104.2822.

13. Lee Smolin, "Precedence and Freedom in Quantum Physics" (2012), arXiv:1205.3707.

Chapter 15: A Causal Theory of Views

1. Luca Bombelli, Joohan Lee, David Meyer, and Rafael D. Sorkin, "Space-Time as a Causal Set," *Physical Review Letters* 59, no. 5 (August 1987): 521–24; Sorkin, "Spacetime and Causal Sets," in *Relativity and Gravitation: Classical and Quantum* (Proceedings of the SILARG VII Conference, held in Cocoyoc, Mexico, December 1990), eds. J. C. D'Olivo et al. (Singapore: World Scientific, 1991), 150–73.
2. Maqbool Ahmed, Scott Dodelson, Patrick B. Greene, and Rafael Sorkin, "Everpresent Lambda," *Physical Review D* 69, no. 10 (May 2004): 103523, arXiv:astro-ph/0209274.
3. Ted Jacobson, "Thermodynamics of Spacetime: The Einstein Equation of State," *Physical Review Letters* 75, no. 7 (August 1995): 1260, arXiv: gr-qc/9504004.
4. Fotini Markopoulou and Lee Smolin, "Holography in a Quantum Spacetime" (October 1999), arXiv:hep-th/9910146; Smolin, "The Strong and Weak Holographic Principles," *Nuclear Physics B* 601, nos. 1–2 (May 2001): 209–47, arXiv:hep-th/0003056.
5. Marina Cortês and Lee Smolin, "The Universe as a Process of Unique Events," *Physical Review D* 90, no. 8 (October 2014): 084007, arXiv:1307.6167 [gr-qc]; Cortês and Smolin, "Quantum Energetic Causal Sets," *Physical Review D* 90, no. 4 (August 2014): 044035, arXiv:1308.2206 [gr-qc]; Cortês and Smolin, "Spin Foam Models as Energetic Causal Sets," *Physical Review D* 93, no. 8 (June 2014): 084039, arXiv:1407.0032; Cortês and Smolin, "Reversing the Irreversible: From Limit Cycles to Emergent Time Symmetry," *Physical Review D* 97, no. 2 (January 2018): 026004, arXiv:1703.09696.
6. Smolin, "The Dynamics of Difference," *Foundations of Physics* 48, no. 2 (2018): 121–34, arXiv:1712.04799.

Epilogue/Revolutions

Epigraph David Gross, "Closing Remarks," Strings 2003 Conference, Kyoto, Japan, July 6–11, 2003, slide 17, https://www.yukawa.kyoto-u.ac.jp/assets/contents/seminar/archive/2003/str2003/talks/gross.pdf.

GLOSSARY

Acceleration: The rate of change of velocity.

Angular momentum: A conserved quantity that measures the amount of rotation or angular motion.

Anti-realism: A philosophy according to which either there is no objective, universal reality, or if there is such, human beings cannot have complete knowledge of it.

Atom: The basic unit of matter, consisting of a nucleus, which contains protons and neutrons, surrounded by electrons.

Background: A scientific model or theory often describes only part of the universe. Some features of the rest of the universe may be included as necessary to define the properties of that part of the universe that is studied. These features are called the background. For example, in Newtonian physics space and time are part of the background because they are taken to be absolute.

Background dependent: A theory, such as Newtonian physics, that makes use of a background.

Background independent: A theory that does not make use of a division of the universe into a part that is modeled and the rest, which is taken to be part of the background. General relativity is said to be background independent because the geometry of space and time is not fixed, but evolves in time just like any other field, such as the electromagnetic field.

Bayesian probability: A subjective probability which measures a person's degree of belief about something.

Bell's theorem: States that in a world which is local, in the sense that the choice of measurements made on one system never influences the probabilities for the outcome of measurements made on a distant system, certain correlations of measurements are restricted by an inequality. That inequality is violated experimentally. Also called Bell's relation or Bell's restriction.

Bohmian mechanics: Another name for pilot wave theory.

Causal set theory: An approach to quantum spacetime based on the hypothesis that the history of the world is made from a discrete set of fundamental events and their causal relations.

Causality: The principle that events are influenced by those in their past. In relativity theory one event can have a causal influence on another only if energy or information sent from the first reaches the second.

Causal structure: Because there is a maximum speed at which energy and information can be transmitted, the events in the history of the universe can be organized in terms of their possible causal relations. To do this, one indicates, for every pair of events, whether the first is in the causal future of the second, or vice versa, or whether there is no possible causal relation between them because no signal could have traveled between them. Such a complete description defines the causal structure of the universe.

Classical physics: That part of physics from Galileo through general relativity, prior to the quantum theory.

Collapse of the wave function: The postulate that immediately after an observer takes a measurement which reveals a definite value for some observable, a quantum system takes on the quantum state associated to that value.

Complementarity principle: Principle proposed by Bohr that quantum systems admit different descriptions, such as particle and wave, that would contradict each other if they had to be imposed simultaneously. However, any given experiment can be described using one or the other.

Conserved quantity: A property of a physical system whose total value never changes in time as the system evolves. Examples are energy, momentum, and angular momentum.

Consistent histories approach: An interpretation of quantum mechanics based on assigning probabilities to sets of histories that decohere from each other.

Contrary state: See Einstein-Podolsky-Rosen state.

De Broglie–Bohm theory: Another name for pilot wave theory, named for its two inventors.

Decoherence: The process by which large quantum systems, containing many degrees of freedom, in contact with an environment which introduces random fluctuations, lose their wave properties, due to the phases of the waves becoming randomized, and so emerge as particles.

Degree of freedom: A variable quantity, describing one way a physical system can change.

Determinism: The philosophy that the future state of a physical system is completely determined by the laws of physics acting on the present state.

Discreteness: The property of some observables of quantum systems, such as the energy of an atom, to take values restricted to a discrete list.

Dynamical collapse theory: A proposal that collapse of the wave function is a real physical process.

Einstein-Podolsky-Rosen (EPR) state: A joint state of two particles which contains no information at all about the individual particles, but indicates that if any measurement is made on both, the results will be opposite. Also called the contrary state.

Energy: A physical quantity giving a measure of the activity of a system, whose value is preserved in time. Energy takes several forms and can be transmuted among them, with the total value always conserved.

Entanglement: A property of a quantum state of two or more systems, where the state indicates a property shared by those systems that is not just the sum of properties held by the individual particles. The EPR or contrary state is an example of an entangled state.

Entropy: A measure of the disorder of a physical system, which is related to the information trapped in the exact values of its microscopic degrees of freedom.

Event: In relativity theory, something that happens at a particular point of space and moment of time.

Exclusion principle: Invented by Wolfgang Pauli, it says that no two fermions can be in the same quantum state.

Field: A physical system spread out in space, with one or more degrees of freedom per spacetime point. The electric and magnetic fields are examples.

Field theory: A physical theory that describes the evolution in time of one or several fields. An example is electrodynamics, where the laws of motion of the fields are called the Maxwell equations.

Force: In Newtonian physics, the change in the momentum in a collision. Also equal to the acceleration of a body times the mass.

Future: The future, or causal future, of an event consists of all those events that it can influence by sending energy or information to them.

Hidden variable: A property or degree of freedom of a quantum system that is not described by quantum mechanics, but is needed to complete the description of an individual system.

Holographic principle: A conjectured principle which limits the quantity of information crossing a surface to the area of the surface in Planck units.

Information: A measure of the organization of a signal. It is equal to the number of yes/no questions whose answers could be coded in the signal.

Instrumentalism: An approach to science wherein the role of theory is only to provide a description of a physical system in terms of its responses to externally imposed forces conveyed by measuring instruments.

Kochen-Specker theorem: A theorem that shows that quantum mechanics is contextual, which means that the value of an observable can depend on a choice of which other measurements are made at the same time.

Locality: The property of physical law that systems are only directly influenced by what is nearby in space and time.

Loop quantum gravity: An approach to quantum gravity based on a quantization of Einstein's general theory of relativity.

Many moments interpretation: The hypothesis that what really exists is a vast collection of moments, containing everything that might have happened in the history of the universe.

Many Worlds Interpretation: An interpretation of quantum theory according to which the different possible outcomes of an observation of a quantum system reside in different universes, all of which somehow coexist.

Mass: In Newtonian physics, the inertial mass is a measure of the quantity of matter, which, multiplied by velocity, gives a conserved quantity called the momentum.

Matrix: A table of numbers organized into rows and columns.

Matrix mechanics: An approach to quantum mechanics in which observables are represented by matrices.

Momentum: A quantity defined for moving particles, which is exchanged in collisions so as to conserve the total. In Newtonian physics it is equal to the product of the mass and velocity.

Newtonian physics: A framework for describing and explaining motion, invented by Isaac Newton and presented in his 1687 book *Principia Mathematica*, which is based on three laws of motion.

Nonlocality: Any phenomenon which does not satisfy the principle of locality, and so involves influences transmitted between systems separated in space.

Operationalism: An approach to instrumentalism in which one specifies for a physical system a set of operations which include how it is to be prepared and how it is to be measured.

Past or causal past: For a particular event, all other events that could have influenced it by sending energy or information to it.

Photon: A quantum of the electromagnetic field, which carries an amount of energy proportional to the frequency of the field.

Pilot wave theory: The first realist approach to quantum mechanics, invented by Louis de Broglie in 1927 and reinvented by David Bohm in 1952. A complete description of an individual system is given by both a wave and a particle, where the particle is guided by the wave.

Planck's constant: The fundamental quantity specifying the scale at which the effects of quantum physics depart from those of Newtonian physics. Usually represented as h. It comes into the relationships between the energy of a quantum and the frequency of the related wave.

Planck energy: A unit of energy constructed by multiplying Planck's constant, h, Newton's gravitational constant, G, and the speed of light, c, together in the right combination to give an energy. It is equal to the energy in one hundred-thousandth of a gram.

Planck length: The unit of length so constructed; it is roughly twenty powers of ten smaller than an atomic nucleus.

Planck mass: The unit of mass so constructed, about one hundred-thousandth of a gram.

Quanta (n., pl.): The particle side of the wave-particle duality.

Quantize (v.): To follow an algorithm that takes as input a classical or Newtonian theory and outputs a corresponding quantum theory. It is known that any such algorithm is highly non-unique.

Quantum Bayesianism: An approach to quantum foundations according to which all uses of probability in quantum mechanics are subjective, betting probabilities.

Quantum cosmology: The theory that attempts to describe the whole universe in the language of quantum theory.

Quantum equilibrium: In a hidden variable theory such as pilot wave theory, the statistical distribution of particles in an ensemble of systems is arbitrary. When it is equal to the square of the wave function, as is specified in Born's rule, the system is said to be in quantum equilibrium.

Quantum field theory: A quantum theory of fields such as the electric and magnetic fields. These are challenging because they must incorporate special relativity and also because they have an infinite number of degrees of freedom.

Quantum gravity: The theory which combines general relativity and quantum physics.

Quantum mechanics: The theory of atoms and light as developed in the 1920s.

Quantum state: A complete description of an individual system according to quantum mechanics.

Realism: The belief that there is an objective physical world whose properties are independent of what human beings know or which experiments we choose to do. Realists also believe that there is no obstacle in principle to our obtaining complete knowledge of this world.

Relationalism: The philosophy that all the properties of elementary objects or events arise from interactions between pairs or larger sets of them, and hence measure shared properties.

Relational quantum theory: An interpretation of quantum theory according to which the quantum state of a particle, or of any subsystem of the universe, is defined not absolutely, but only in a context created by the presence of an observer, and by a division of the universe into a part containing the observer and a part containing that part of the universe from which the observer can receive information. Relational quantum cosmology is an approach to quantum cosmology which asserts that there is not one quantum state of the universe, but as many states as there are such contexts.

Relativity, the general theory of: Einstein's 1915 theory of gravitation in which the gravitational force is replaced by the dynamics of the spacetime geometry.

Relativity, the special theory of: Einstein's 1905 theory of motion and light in the absence of gravity.

Retrocausality: Hypothetical processes in which the order of causes runs backward relative to the global direction of time.

Rule 0: The basic dynamical equation of quantum gravity, which expresses the absence of a global or universal time. Also called the Wheeler-DeWitt equation.

Rule 1: The basic dynamical equation of quantum mechanics that describes how quantum states evolve with respect to time as measured by clocks outside the quantum system. Also called the Schrödinger equation. Rule 1 explains that given the quantum state of an isolated system at one time, there is a law that will predict the precise quantum state of that system at any other time.

Rule 2: The law that prescribes how a quantum state responds to a measurement, which is to collapse immediately into a state within which the measured quantity has a precise value, the value that the measurement produced. Rule 2 explains that the outcome of a measurement can only be predicted probabilistically. But afterward, the measurement changes the quantum state of the system being measured, by putting it in the state corresponding to the result of the measurement. This is called collapse of the wave function.

Schrödinger's cat experiment: A thought experiment in which Rule 1 implies that a cat is in a superposition of two distinct macroscopic states: living and dead.

Schrödinger's equation: See Rule 1.

Second law of thermodynamics: States that the entropy of an isolated system will most probably increase.

Speed: The rate of change of distance with time.

Spin: The angular momentum of an elementary particle which is an intrinsic property of it, independent of its motion.

Spin network: A graph whose edges are labeled by numbers representing spins. In loop quantum gravity each quantum state of the geometry of space is represented by a spin network.

Standard model of particle physics: A quantum field theory which is our best model of the elementary particles and their interactions, except for gravity.

State: In any physical theory, the configuration of a system at a specified moment of time.

String theory: An approach to quantum gravity based on the hypothesis that the fundamental things in the world are one-dimensional.

Symmetry: An operation by which a physical system may be transformed without affecting the fact that its state is a possible state of the system. Two states connected by a symmetry have the same energy.

Uncertainty principle: A principle in quantum theory according to which it is impossible to measure both the position and momentum (or velocity) of a particle.

Velocity: The rate of change of position in time.

Wave function: A representation of the quantum state of a system.

Wave mechanics: A form of quantum mechanics invented by Erwin Schrödinger in 1926. Later shown to be equivalent to matrix mechanics.

Wave-particle duality: A principle of quantum theory according to which one can describe elementary particles as both particles and waves, depending on the context.

FURTHER READING

Popular Books by the Inventors of Quantum Mechanics

Bell, J. S. *Speakable and Unspeakable in Quantum Mechanics.* 2nd ed. Introduction by Alain Aspect; two additional papers. Cambridge, UK: Cambridge University Press, 2004.

Bohm, David. *Wholeness and the Implicate Order.* London: Routledge and Kegan Paul, 1980. Reprint, London: Ark / Routledge, 2002.

Bohr, Niels. *Atomic Physics and Human Knowledge.* New York: Science Editions, 1961. Reprint, Mineola, NY: Dover Publications, 2010.

Bohr, Niels. *Atomic Theory and the Description of Nature: Four Essays with an Introductory Survey.* Cambridge, UK: Cambridge University Press, 1934, 1961. Reprint, 2011.

Bohr, Niels. "Discussion with Einstein on Epistemological Problems in Atomic Physics." In *Albert Einstein: Philosopher-Scientist,* edited by Paul Arthur Schilpp, 199–242. 3rd ed. Library of Living Philosophers 7. Peru, IL: Open Court Publishing, 1988.

Einstein, Albert. *Autobiographical Notes.* Translated and edited by Paul Arthur Schilpp. Centennial ed. Peru, IL: Open Court Publishing, 1999.

Einstein, Albert. *Ideas and Opinions.* Reprint ed. New York: Broadway Books, 1995.

Heisenberg, Werner. *Philosophical Problems of Quantum Physics.* 2nd ed. Woodbridge, CT: Ox Bow Press, 1979.

Heisenberg, Werner. *The Physical Principles of the Quantum Theory.* Translated by Carl Eckart and F. C. Hoyt. Mineola, NY: Dover Publications, 1949.

Schrödinger, Erwin. *What Is Life? With Mind and Matter* and *Autobiographical Sketches*. Foreword to *What Is Life?* by Roger Penrose. Cambridge, UK: Canto / Cambridge University Press, 1992.

Books by Contemporary Contributors

Barbour, Julian. *The End of Time: The Next Revolution in Our Understanding of the Universe*. New York: Oxford University Press, 1999.

Carroll, Sean. *The Big Picture: On the Origins of Life, Meaning, and the Universe Itself.* New York: Dutton, 2016.

Deutsch, David. *The Beginning of Infinity: Explanations that Transform the World*. New York: Viking, 2011.

Deutsch, David. *The Fabric of Reality: The Science of Parallel Universes—and Its Implications*. New York: Penguin Press, 1997.

Greene, Brian. *The Hidden Reality: Parallel Universes and the Deep Laws of the Cosmos*. New York: Alfred A. Knopf, 2011.

Penrose, Roger. *The Emperor's New Mind: Concerning Computers, Minds, and The Laws of Physics*. Reprint ed., with a new preface by the author. Oxford and New York: Oxford University Press, 1999.

Penrose, Roger. *Shadows of the Mind: A Search for the Missing Science of Consciousness.* Oxford and New York: Oxford University Press, 1994.

Rovelli, Carlo. *The Order of Time*. New York: Riverhead Books, 2018. // *L'ordine del tempo*. Milan: Adelphi Edizioni, 2017.

Rovelli, Carlo. *Reality Is Not What It Seems: The Journey to Quantum Gravity*. New York: Riverhead Books, 2017. // *La realtà non è come ci appare: La struttura elementare delle cose*. Milan: Raffaello Cortina Editore, 2014.

Rovelli, Carlo. *Seven Brief Lessons on Physics*. New York: Riverhead Books, 2016. // *Sette brevi lezioni di fisica*. Milan: Adelphi Edizioni, 2014.

Tegmark, Max. *Our Mathematical Universe: My Quest for the Ultimate Nature of Reality*. New York: Alfred A. Knopf, 2014.

Biographies of Key Figures

Byrne, Peter. *The Many Worlds of Hugh Everett III: Multiple Universes, Mutual Assured Destruction, and the Meltdown of a Nuclear Family*. Oxford and New York: Oxford University Press, 2010.

Farmelo, Graham. *The Strangest Man: The Hidden Life of Paul Dirac, Mystic of the Atom*. New York: Basic Books, 2009.

Gribbin, John. *Erwin Schrödinger and the Quantum Revolution*. Hoboken, NJ: John Wiley and Sons, 2013.

Hoffmann, Banesh, with Helen Dukas. *Albert Einstein: Creator and Rebel*. New York: Viking Press, 1973.

Klein, Martin J. *Paul Ehrenfest. Vol. 1: The Making of a Theoretical Physicist*. New York: American Elsevier, 1970.

Overbye, Dennis. *Einstein in Love: A Scientific Romance*. New York: Penguin, 2000.

Pais, Abraham. *Niels Bohr's Times: In Physics, Philosophy, and Polity*. Oxford, UK, and New York: Clarendon Press / Oxford University Press, 1991.

Pais, Abraham. *Subtle is the Lord: The Science and the Life of Albert Einstein*. Oxford, UK, and New York: Oxford University Press, 1982. Reprint ed., with a new foreword by Roger Penrose, 2005.

Peat, F. David. *Infinite Potential: The Life and Times of David Bohm*. Reading, MA: Addison-Wesley, 1997.

Histories of Quantum Physics

Bacciagaluppi, Guido, and Antony Valentini. *Quantum Theory at the Crossroads: Reconsidering the 1927 Solvay Conference*. Cambridge, UK, and New York: Cambridge University Press, 2009.

Baggott, Jim. *The Quantum Story: A History in 40 Moments*. Oxford, UK, and New York: Oxford University Press, 2011.

Baggott, Jim. *Beyond Measure: Modern Physics, Philosophy, and the Meaning of Quantum Theory*. Oxford, UK, and New York: Oxford University Press, 2004.

Forman, Paul. "Weimar Culture, Causality, and Quantum Theory, 1918–1927: Adaptation by German Physicists and Mathematicians to a Hostile Intellectual Environment." *Historical Studies in the Physical Sciences*, Vol. 3 (1971): 1–115. Forman expanded on his original argument in: Forman, Paul. "*Kausalität*, *Anschaulichkeit*, and *Individualität*, or How Cultural Values Prescribed the Character and the Lessons Ascribed to Quantum Mechanics." In *Society and Knowledge: Contemporary Perspectives in the Sociology of Knowledge and Science*, edited by Nico Stehr and Volker Meja, 333–47. New Brunswick, NJ: Transaction Books, 1984.

Gefter, Amanda. *Trespassing on Einstein's Lawn: A Father, a Daughter, the Meaning of Nothing, and the Beginning of Everything.* New York: Bantam Books, 2014.

Gilder, Louisa. *The Age of Entanglement: When Quantum Physics Was Reborn.* New York: Alfred A. Knopf, 2008.

Gribbin, John. *In Search of Schrödinger's Cat: Quantum Physics and Reality.* New York: Bantam Books, 1984.

Jammer, Max. *The Philosophy of Quantum Mechanics: The Interpretations of Quantum Mechanics in Historical Perspective.* New York: John Wiley and Sons, 1974.

Kaiser, David. *How the Hippies Saved Physics: Science, Counterculture, and the Quantum Revival.* New York: W. W. Norton, 2011.

Kragh, Helge. *Quantum Generations: A History of Physics in the Twentieth Century.* Princeton: Princeton University Press, 1999. Reprint, 2002.

Kuhn, Thomas S. *Black-Body Theory and the Quantum Discontinuity, 1894–1912.* Chicago: University of Chicago Press, 1987.

Stone, A. Douglas. *Einstein and the Quantum: The Quest of the Valiant Swabian.* Princeton: Princeton University Press, 2013.

Collections of Papers

DeWitt, Bryce Seligman, and Neill Graham, eds. *The Many Worlds Interpretation of Quantum Mechanics.* Princeton Series in Physics. Princeton: Princeton University Press, 1973. Reprint ed.: Princeton Legacy Library, 2015.

Saunders, Simon, Jonathan Barrett, Adrian Kent, and David Wallace, eds. *Many Worlds? Everett, Quantum Theory, and Reality.* Oxford: Oxford University Press, 2010.

INDEX

INDEX